HETERICK MEMORIAL LIBRARY
OHIO NORTHERN UNIVERSITY
ADA, OHIO 45810

Japan's Technical Standards

Implications for Global Trade and Competitiveness

Edited by
JOHN R. McINTYRE

QUORUM BOOKS
Westport, Connecticut • London

Library of Congress Cataloging-in-Publication Data

Japan's technical standards : implications for global trade and
  competitiveness / edited by John R. McIntyre.
        p.   cm.
    Includes bibliographical references and index.
    ISBN 1-56720-053-2 (alk. paper)
    1. Standardization—Japan. 2. Technology and state—Japan.
  I. McIntyre, John R.
  T59.2.J3J38   1997
  332'.02'021852—dc20   96-2212

British Library Cataloguing in Publication Data is available.

Copyright © 1997 by John R. McIntyre

All rights reserved. No portion of this book may be
reproduced, by any process or technique, without the
express written consent of the publisher.

Library of Congress Catalog Card Number: 96-2212
ISBN: 1-56720-053-2

First published in 1997

Quorum Books, 88 Post Road West, Westport, CT 06881
An imprint of Greenwood Publishing Group, Inc.

Printed in the United States of America

The paper used in this book complies with the
Permanent Paper Standard issued by the National
Information Standards Organization (Z39.48–1984).

10 9 8 7 6 5 4 3 2 1

Contents

Tables and Figures		ix
Acronyms		xi
Foreword		
Robert G. Hawkins		xv
Introduction: Standard Setting in Japan—Trends and Issues		
John R. McIntyre		xvii

PART I   **The Japanese Technoinfrastructure**

1   Technological Infrastructure and the Role of Government: Comparative Perspectives
      Gregory Tassey   3

2   Regional Technology Infrastructures for Japanese Small- and Medium-Sized Enterprises: Policies and Programs for Technological Modernization
      Philip P. Shapira   27

3   Living on the Edge: Basic Research and Knowledge Creation in Japanese Electronics Companies
      David T. Methé   47

PART II:	The Japanese Technical Standards System and Techno-Competition in Select Japanese Industries	
4	Globalization and the Role of Standards *Koji Tanabe*	67
5	The Japanese Technology Infrastructure: Issues and Opportunities *John P. Stern*	75
6	Technical Standards and Access to the Japanese Cellular Communications Equipment Market *Douglas J. Puffert*	87
7	Introducing a Standard for Digitalized Medical Images in Japan *Aki Yoshikawa*	93
8	Critical Success Factors Leading to Japanese Dominance in Electronics Packaging *Michael J. Kelly*	107
PART III:	International Trade, Technical Standards, and Barriers in Japan	
9	The Evolution of Technical Standards and Trade in a Changing World Economy *Stanley I. Warshaw*	119
10	Technical Regulations and Trade: New Developments and the Asia Pacific Economic Cooperation Forum *John Sullivan Wilson*	125
11	Japan's Double Standards: Technical Standards and U.S.–Japan Economic Relations *Brian Woodall*	145
12	The Uruguay Round's Agreement on Technical Barriers to Trade: Implications for the United States and Japan *William J. Long and Kimberly Wildner*	163

Glossary of Japanese Terms 175
Index 177
About the Editor and Contributors 183

Tables and Figures

TABLES

1.1	Examples of Technology-Based Market Failures and Government Policy Responses	12
2.1	Small- and Mid-Sized Enterprises, Japan 1991	28
2.2	Japan's Public Technology Centers Measures of Performance	36
2.3	Structure and Organization of Japan's Kohsetsushi and Regional Technology Centers	42
3.1	Number of New Product Introductions, Selected Products, 1987–1991	50
3.2	Number of Basic Research Institutes Established in Japan	54
3.3	Various Approaches Taken toward Developing the Knowledge Exploration Capabilities of Ten Leading Japanese Electronic Companies	55
5.1	Who's Really on Top?	76
5.2	1994 Japan–U.S. Electronics Trade Top Exports-Imports	78
5.3	ISO 9000 in Japan: A 1992 Survey	82
5.4	Commercialization and Standardization in Japan	84
6.1	Global Cellular Communications System Standards	89

FIGURES

1.1	Public-to-Private Capital Stock Ratio	5
1.2	Disaggregated Technology Growth Model	9

2.1	Small Manufacturing Establishments and Employment, Japan and United States	29
2.2	Japan's Public Support System for SME Technology Promotion	34
2.3	Kohsetsushi Centers and Regional Technology Projects in Japan	35
2.4	Kohsetsushi Centers Technological Specialties and Size of Staff	37
2.5	Research Core Concept	40
3.1	Comparison of Knowledge Exploration and Knowledge Exploitation Approaches toward Research and Development in Companies	49
3.2	NEC PC-9800 Product Family Group	51
3.3	Integrated versus Insulated Spectrum	62
5.1	Percentage of Real Growth in the Standard of Living, 1973–1993	79
5.2	JIS Standardization Procedures and Policies to Guarantee Their Transparency	81
5.3	Mastering Japanese Technical Standards	85
7.1	International Comparison of CT Diffusion	94
7.2	CT Market in Japan	95
7.3	MRI Market in Japan	95
7.4	Diffusion of CR in Japan	96
7.5	PACS in Japan	97
7.6	IS&C Concept	100
8.1	Mass Production Strategy for Low-Cost Electronic Packaging Advances	109

Acronyms

A2LA	American Association for Laboratory Accreditation
ABS	American Bureau of Shipping (Industrial Verification)
ACR	American College of Radiology
AIST	Agency for Industrial Science and Technology (Japan)
AMPS	Advanced Mobile Phone Service
ANSI	American National Standards Institute
APEC	Asia Pacific Economic Corporation
APLAC	Asia-Pacific Laboratory Accreditation Cooperation
APLMF	Asia-Pacific Legal Metrology Forum
APMP	Asia-Pacific Metrology Program
ARL	Advanced Research Laboratory
ATP	Advanced Technology Program
BOF	Basic Oxygen Furnace
BRI	Basic Research Institute
BRL	Basic Research Laboratory
CAD	Computer Aided Design
CAM	Computer Aided Manufacturing
CDMA	Code Division Multiple Access
CEN	European Committee for Standardization (Comité Européen de Normalisation)
CENELEC	European Committee for Electrotechnical Standardization Center
CFTA	U.S.–Canada Free Trade Agreement
CIAJ	Communications Industry Association of Japan
CR	Computed Radiography

DRAM	Dynamic RAM
DSB	Dispute Settlement Body
DSU	Understanding on Rules and Procedures Governing the Settlement of Disputes
DVD	Digital Video Disk
ESPRIT	European Strategic Program for R&D in Information Technology
ETSI	European Telecommunications Standards Institute
EU	European Union
FCC	Federal Communications Committee (U.S.)
FDD	Floppy Disk Drives
FILP	Fiscal Investment and Loan Program
FQA	Fastener Quality Act (1990)
FSU	Former Soviet Union
FTAA	Free Trade Area of the Americas
GATT	General Agreement on Tariffs and Trade
GDP	Gross Domestic Product
GNP	Gross National Product
GIRI	Government Industrial Research Institutes
GSM	Global System for Mobile Communications (EU)
HIS	Hospital Information Systems
IC	Integrated circuit
IEC	International Electrotechnical Committee
IGC	Intergovernmental Council for Standardization Certification and Metrology
ILO	International Labour Office
IMS	Intelligent Manufacturing System
IPR	Intellectual Property Rights
IS&C	Image Save and Carry
ISO	International Standards Organization
ITU	International Telecommunication Union
JESSI	Joint European Submicron Silicon Initiative
JETRO	Japan External Trade Organization
JIRA	Japan Industries of Radiation Apparatus
JIS	Japanese Industrial Standards
JISC	Japanese Industrial Standards Committee
JRDC	Research and Development Corporation of Japan
JSA	Japan Standards Association
JSS	Japan Standards Systems
JTEC	Japanese Technology Evaluation
KTC	Key Technology Center
MEDIS	Medical Information System Development Center
MELCO	Mitsubishi Electric
MHW	Ministry of Health and Welfare (Japan)
MIPS	Medical Imaging and Processing Systems
MITI	Ministry of International Trade and Industry
MOD	Magneto Optical Disks

Acronyms xiii

MOS	Metal Oxide Semiconductor
MOSS	Market Oriented Sector Service (Agreement)
MPT	Ministry of Posts and Telecommunications (Japan)
MRA	Mutual Recognition Agreements
MRI	Magnetic Resonance Imaging
MSTQ	Metrology, Standards, Testing and Quality
NAFTA	North American Free Trade Agreement
NEDO	New Energy and Industrial Development Organization
NEMA	National Electrical Manufacturers' Association
NIS	Newly Independent States
NIST	National Institute for Standards and Technology
NMT	Nordic Mobile Telephone
NTT	Nippon Telephone and Telegraph
NVCASE	National Voluntary Conformity Assessment Systems Evaluation
NVLAP	National Voluntary Laboratory Accreditation Program
OEM	Original Equipment Manufacturing
OTA	Office Technology Assessment
OTO	Office of the Trade and Investment Ombudsman
PAC	Pacific Accreditation Cooperation
PACS	Picture Archiving and Communication Systems
PASC	Pacific Area Standards Congress
PNGV	Partnership for a New Generation Vehicle
PWB	Printed wiring board
RoR	Rate of Return
RWCS	Real World Computer System
SASO	Saudi Arabia Standards Organization
SCSC	Sub-Committee on Standards and Conformance
SII	Structural Impediments Initiative
SME	Small- and Medium-Sized Manufacturing Enterprises
SMEA	Small and Medium Enterprise Agency
SMT	Surface mount technologies
SPS	Sanitary and Phytosanitary Standards
SRM	Standards Related Matters
STA	Science and Technology Agency
TACS	Total Access Communications System (Motorola)
TBT	Technical Barriers to Trade
TDMA	Time Division Multiple Access (US)
TQM	Total Quality Management
UL	Underwriters Laboratory
UNEP	United Nations Environment Programme
UR	Uruguay Round
USTC	United States Technical Committee
USTR	United States Trade Representative
WHO	World Health Organization
WTO	World Trade Organization

Foreword

Robert G. Hawkins

The use of product standards, specifications and certifications as a means of denying market access to foreign suppliers is the most important issue in international merchandise trade policy today. As recently as fifteen years ago, it was only one among many "non-tariff barriers" which had emerged as tariff rates were negotiated downward on most products from 1945 to 1980. Since 1980, a multitude of non-tariff barriers have been negotiated away, or unilaterally eliminated by countries, the most dramatic example being the internal trade of the European Union.

But, like international protection for intellectual property (patent, copyright, and trademark piracy) in international trade in services, "product standards" as a barrier to international trade in merchandise (goods) has emerged as a sticky, almost intractable issue—jealously protected by importing countries and their companies; aggressively pursued by potential exporters. No place is this more conflictive than in Japan-United States trade.

There are several reasons why this issue is so touchy and difficult. One is that product standards and specifications have a legitimate place in national policies to protect consumers and customers, and to impose common specifications for components or performance to permit interchangeability among products of different suppliers. Often, these specifications are set by industry groups; in other instances by government bodies.

At the same time, such product standards and specifications may be stated in a way as to exclude foreign suppliers from competition—to provide competitive advantage to local suppliers. This is even more problematic when the certification of the products of a particular foreign supplier must be made by agents of the importing country or through some other means, and when it must be performed by individual units or through plant or process approval.

All of this becomes more complex as multinational firms produce components and final products in several foreign locations, as well as at home. Is the imported Nissan model to Japan, from a British subsidiary, a British car, or a Japanese car, and who certifies that it (and its components) meets Japanese standards.

This volume addresses these issues—from several levels. One is at the national level; i.e. how are individual nations or groups of nations dealing with the product standard issues. Another is the individual company or industry. And a third is the international mechanisms for policy negotiations and for standard setting. In the process, the perspectives covered are those of the imposing countries, the companies which attempt to comply, and the nations which seek to expand experts.

We trust that this volume makes a contribution to our understanding of this complex issue, and appreciate the efforts of the contributors, and the editorial dedication of the Georgia Tech Center for International Business and Education and Research staff, in particular Ms. Sandra Beaudin, CIBER graduate assistant, who took the task of turning the raw material into finished text.

Introduction: Standard Setting in Japan—Trends and Issues

John R. McIntyre

This book is the result of a major research conference held in Atlanta, Georgia, at the Georgia Tech Manufacturing Research Center on May 19, 1995, and of a follow-up business symposium held at the Hotel Nikko-Atlanta on June 9, 1995. These two events brought together technical-standards experts from the Japanese and U.S. academic, business, and government communities. Twelve papers from a variety of disciplinary perspectives (ranging from manufacturing to marketing, from law to political science) were commissioned, presented, and discussed. These two events were made possible with the generous support of the U.S. Department of Education, under a Center for International Business Education grant to the Georgia Institute of Technology, the U.S. Department of Commerce's Japan Technology Program, the National Institute of Standards and Technology, as well as the Atlanta office of the Japan External Trade Organization (JETRO).

A number of policy questions are addressed in this volume:

- How does the Japanese technical standards system work in a national and comparative light?
- How do Japanese technical standards impact on market penetration strategy and market share growth by non-Japanese firms?
- How do they affect well-established Japanese trading successes in given sectors?
- How do Japanese technical standards fit in the overall Japanese "technoinfrastructure" in explaining Japanese global R&D competitiveness?

Answers to these questions are of keen interest to business decision makers and government policy makers. While a growing body of research literature bearing on the economics of technical standards and harmonization of standards in the European Union has emerged in the past fifteen years, many gaps in our knowledge and understanding of these topics remain. This is particularly so as regards Japan and the role of technical standards in its competitiveness and industrial policy. This gap is egregious given the importance of Japan in the world economy and its share of world trade and manufacturing investment. This knowledge gap is further accentuated by the fact that most of the relevant literature is in Japanese, and that Western scholars have insufficiently focused on this aspect of Japanese industrial structure and competitiveness.

This is the first book-size, systematic treatment of Japanese technical standards and their impact on trade, industrial policy, and competitiveness in the English language. While it makes no claim to definitive answers to the questions it raises, it provides a framework and suggests further research directions as well as prescriptive policies.

IN SEARCH OF AN ELUSIVE DEFINITION

There is no single and simple definition of what a standard is that captures the full range of meanings. For purposes of this work, a standard is a set of characteristics or quantities that describes features of a product, process, service, interface, or material (National Research Council, 1995, p. 9). It can be formal or informal as a social custom is. A formal standard, however, is designed to meet a specific objective and is specified by its developers so as to make it usable. The literature distinguishes product/quality standards (e.g., VCR standards, NSA encryption standards, refrigerator standards), control standards (e.g., database-privacy standards, auto safety and fuel economy standards, pressure vessel standards), and process/interoperability standards (e.g., computer interface standards, electronic data interexchange standards, open network architecture standards).

The literature further distinguishes three different methods of setting standards. First, formal standards can be written unilaterally by a product designer, manufacturer, or purchaser. In this case, the standard is set through the market on a de facto basis. Second, they can be negotiated through cooperation and consensus among a group of concerned parties. Third, they can be mandated by government through a regulatory process. Often the standard setting procedure moves on several fronts at the same time. An industry group may endorse a standard developed unilaterally by one of its members. A government agency may then adopt the standard through regulation.

THE UNITED STATES AND INDUSTRIALIZED COUNTRIES: CONTRASTING APPROACHES

In the United States, almost half of all standards are set by the private sector as part of a voluntary consensual process. It reflects core American political values and the general preference for a market-based, pluralistic approach. R. B. Toth has described the U.S. standardization system as "distributed, pragmatic, reactionary, entrepreneurial and individualistic," and endowed with a great amount of "openness and transparency" as well as "self-certification and warranties," clear appeals procedures, implementation often challenged, and one in which international standards only fulfill the function of guidance rather than regulation (Office of Technology Assessment, March 1992, p. 14).

In contrast, Toth has characterized other industrialized nations' standard setting systems as endowed with the following characteristics: centralized, systematic, anticipatory, a tool of industrial policy, responsive to government direction and national policy, with fairly immediate implementation and acceptance, infrequent appeals procedures, and third-party testing. He also notes that often these systems directly incorporate international standards into their national standard setting process.

The process of standard development assumes many forms domestically and internationally (for a detailed review of these mechanisms, see National Research Council, 1995, pp. 23–64). This book seeks to shed light on the Japanese process of standards determination. This brief introductory essay provides an overview of the major directions of the recent and extant literature and seeks to position the contributions in this evolving research framework.

STANDARDS, COMPATIBILITY, R&D, AND COMPETITION

Compatibility issues have long been central to industrial economies. The most often-cited, early example is perhaps that of railroad gauges and the use of interchangeable parts as an important step in the industrial revolution. The convergence of the computer and telecommunications technologies have made the issue of compatibility an even more important one.

An extensive economics-based literature has been developed on the relation of compatibility standards and innovation (see for example Choi, 1993; Farrell and Saloner, 1985, 1986). Most standard setting literature is theoretical. It seeks to identify the conditions under which "optimal" standards might emerge. The literature is aimed at the microlevel of the firm and views the producer or vendor as the primary actors in the standards development process.

More recently, the literature has begun to focus on the interrelationship between standards compatibility and competitive strategy (Gabel, 1987). Compatibility is viewed as the result of coordinated product design. The literature has focused on four types of benefit flowing from compatibility:

network externalities (e.g., phone networks and computer networks), competitive effects (i.e., competition is more focused on price and less on design), variety or mix-and-match benefits (e.g., combining different components in a stereo system), and cost savings (i.e., greater economies of scale by reducing the production costs).

In the past, standards were often set only after a product had been fully developed. Nowadays, in the face of rapid technological changes and compressed product life cycles, standards are often set before a product is fully developed. These anticipatory standards are intended to improve the strategic and competitive position of globalized firms (Jensen and Thursby, 1994).

The structure of the market and the interests of stakeholders define, to a large extent, the choice of the standard setting mechanism. If a dominant producer or supplier exists, standards tend to be set in a de facto, market-driven manner. When several manufacturers or suppliers share economic leverage, a process of compromise and negotiation is set in motion. If compelling public interests are at stake, and environmental or safety concerns are involved, governments are likely to become involved (Garvin, 1983).

In relation to the process of innovation, Link and Tassey have identified what they have termed "infratechnologies" (1987, p. 18). They define them as "technologies that support R&D, production and marketing in industries." Infratechnologies include "evaluated scientific data used in the conduct of R&D, measurement and test methods used in research, production, control, and acceptance testing for market transactions; and various technical procedures such as those used in the calibration of equipment." Infratechnologies provide techniques that "enable higher quality and greater reliability at lower cost in production" and "provide buyers and sellers mutually acceptable, low-cost methods of assuring that specified performance levels are met when technologically sophisticated products enter the marketplace" (Ibid., pp. 18–19).

The National Research Council has systematically addressed the functions of standards and their relationships to the process of economic development. Seven key functions for technical standards were identified: enhancement of commercial communication (i.e., reducing the transaction costs between buyers and sellers), facilitation of technology diffusion, improving productive efficiency, heightening of competition (i.e., when products conform to a standard, comparison is easier and competition sharper), easing of interoperability and compatibility, encouraging of continuous internal quality improvement through process management, and protecting public welfare (National Research Council, 1995, pp. 11–17).

TECHNICAL STANDARDS AND INTERNATIONAL TRADE

The economics literature has sought to examine compatibility in an international context. Kende (1991) as well as other treatments of international standards issues (Copeland, 1992; Pelkmans and Beuter, 1978; Hazard and Daems, 1988) have all discussed the trade inhibiting effects of de facto standards, which they have often characterized as a "prohibitive quota." Other writers have also looked at patent races, R&D competition, and their impact on international trade competitiveness in a related literature stream (see Bagwell and Staiger, 1989; Beath, 1990).

The benefits from standardization and compatibility are evidently not limited by national boundaries. Standard choices impact directly on the nature of firm competition and therefore have evident consequences for international trade. Crane was one of the early scholars to look at the implications of international standards on international politics, market transactions, and competition (1979).

An early start in setting standards, or a protected market through government-mandated standards, may lead to a lasting competitive advantage. Standards can also become effective and subtle nontariff barriers. This book addresses this set of issues in some depth with specific reference to Japan.

International standard-setting is often a two-step affair. First, national interests are represented in international standards setting bodies. The more entrenched a nation is in one given standard, the more leverage it has in getting the standard internationally recognized. This may encourage a rush to setting standards prematurely before the innovation process has fully worked itself out. If a national standard fails to prevail internationally, the international market loss can be significant. The stakes are therefore high.

The international trading system has seen major changes in the past few years directly related to international standards. The Uruguay Round of the General Agreement on Tariffs and Trade has resulted in substantial revisions of the Technical Barriers to Trade provisions and the creation of a new section on Sanitary and Phytosanitary Standards for food and agricultural products. Unfair trading practices based on standards, testing, and certification requirements are now prohibited under these two sections of the agreement. The Uruguay Round also addresses the governments' product approval regulations and calls for inclusion of process and production method regulations as well as increased transparency to foreign firms of national and regional standards developing activities. It also seeks to promote mutual acceptance among trading partners of product test results, certification, and conformity assessment measures.

The European Union has taken the lead, as a regional economic bloc, in seeking to harmonize standards among its member states and ultimately in prescribing international standards. Pointing to the active role that governments play in international standards setting bodies has encouraged many to call on the U.S. government to assume greater responsibility in protecting U.S. interests. In fact, a 1990 U.S. International Trade Commission study noted that greater American attention to the issue of standards was a function of European progress in this area. The study stated:

The EC's systematic updating of technical regulations posed the prospect that standards developed as part of the 1992 (Single Market) program might become de facto or de jure world standards. Some claimed that state-of-the-art standards being developed in areas like machine tools could have given European competitors an upper hand, not only in the EC, but in third country markets. (pp. 6–13)

As a result, the United States is in the process of negotiating a number of agreements with its trading partners for mutual recognition of conformity assessment procedures. Mutual recognition will make for the full realization of the trade enhancement aspects of the Uruguay Round and the new World Trade Organization. A noteworthy regional initiative was the establishment, at the behest of the United States, of the Asia Pacific Economic Cooperation Forum (APEC) whose work has focused intensely on technical standards harmonization among its members.

ORGANIZATION OF THE VOLUME

The volume is divided in three parts. Part I (The Japanese Technoinfrastructure) considers the role of government in the process of technology development from three perspectives and three different levels of analysis. Gregory Tassey provides a comparative view of the role of government in the "technological infrastructure," based on the seminal work of Link and Tassey, identified earlier. It offers an analytical framework in which technical standards can be understood as an intrinsic part of the technology development process and of the role played by government. In this regard, it compares the United States, the European Union, and Japan. Philip Shapira focuses on the role of regional and local economic and technology development entities in Japan and the part they have played in technology development. David T. Methé takes his analysis to the firm level by analyzing basic research and knowledge creation in Japanese electronics companies. These three chapters are interdependent, providing insights into Japanese technology development strategies at three different levels and setting the groundwork for a detailed consideration of where and how standards fit in the overall technology development architecture.

Part II (The Japanese Technical Standards System and Technocompetition in Select Japanese Industries) focuses squarely on the mechanics and dynamics of the Japanese technical standards system and the role standards play in making select Japanese industries globally competitive. Koji Tanabe, from an insider's vantage point, provides an overview of evolving trends in the Japanese technical standards system in a context of globalization and harmonization. John Stern presents a unique American perspective, from the vantage point of his Tokyo-based position, on the biases and strengths built into the Japanese system. Douglas Puffert addresses standards development in a specific industry segment—cellular communications equipment, while Aki Yoshikawa focuses on the specific case of digitalized medical imaging. Michael Kelly, drawing on an earlier book-length study, provides insights into Japanese electronics packaging.

Part III (International Trade, Technical Standards, and Barriers in Japan) seeks to assess the impact of the Japanese system on trade competitiveness, fair trade, and whether or not the system evidences some of the characteristics of nontariff barriers. Stanley Warshaw reviews regional efforts at harmonizing standards and how they variously impact on reducing nontariff barriers. John S. Wilson focuses on policy developments in APEC (the Asia Pacific Economic Cooperation Forum) and how this regional initiative can provide a mechanism to reduce standards-related conflicts and harmonize standards as well as certification procedures. Brian Woodall squarely addresses the issue of how Japanese standards often act as nontariff barriers and how this is reflected in U.S.–Japan economic relations. Finally, William Long and Kimberly Wildner assess the impact of the Uruguay Round's Technical Barriers to Trade Code on trade relations between Japan and the United States.

REFERENCES

Bagwell, K., and Staiger, R. "The Sensitivity of R&D Policy to Market Conditions." Mimeo, Stanford University, 1989.

Beath, J. "Models of Technological Competition for the Analysis of Intellectual Property Rights and the Uruguay Round. " Mimeo, 1990.

Choi. "Standardization and Experimentation: Ex Ante vs. Ex Post Standardization." Mimeo, 1993.

Copeland, B. "Product Standards as Trade Policy in an International Duopoly." Mimeo, University of British Columbia, 1992.

Crane, R.J. The Politics of International Standards. Norwood, N.J.: Ablex Publishing, 1979.

Farrell, J., and G. Saloner. "Standardization, Compatibilty, and Innovation. " RAND Journal of Economics, 16, 1985, pp. 70–83.

———. "Installed Base and Compatibility: Innovation, Product Preannouncements, and Predation." American Economic Review, 1986, pp. 940–955.

Gabel, H. Landis, ed. Product Standardization and Competitive Strategy. Amsterdam, Netherlands: Elsevier Science Publishers B.V., 1987.

Garvin, David A. "Can Industry Self-Regulation Work?" California Management Review, Vol. 25, No. 4, Summer 1983.

Hazard, H., and H. Daems. "Technical Standards and Competitive Advantage in World Trade." Mimeo, Harvard University, 1988.

Jensen, R., and M. Thursby. "Patent Races, Product Standards and International Competition." Purdue CIBER, Working Paper 94-015, 1994, Purdue, Indiana.

Kende, M. "Strategic Standardization in Trade with Network Externalities." Mimeo, MIT, 1991.

Link, A. N., and G. Tassey. Strategies for Technology-Based Competition, Meeting the New Global Challenge. Lexington, Mass.: Lexington Books, 1987.

National Research Council. Standards, Conformity Assessment, and Trade. Washington, D.C.: National Academy Press, 1995.

Pelkmans, J., and R. Beuter. "Standardization and Competitiveness: Private and Public Strategies in the EC Color TV industry." In H. Landis Gabel, ed., Op. Cit.

U.S. Congress, Office of Technology Assessment. Global Standards: Building Blocks for the Future. Washington, D.C.: U.S. G.P.O., March 1992.

U.S. International Trade Commission. The Effects of Greater Economic Integration Within the European Community on the United States: First Followup Report. Washington, D.C.: US G.P.O., March 1990.

Part I

THE JAPANESE TECHNOINFRASTRUCTURE

1

Technological Infrastructure and the Role of Government: Comparative Perspectives

Gregory Tassey

Economic research has demonstrated the importance of technology to a modern economy, but it also indicates that significant underinvestment can occur in this critical economic asset (Tassey, 1995). The underinvestment phenomenon results from a number of persistent market "imperfections" or "failures" that lower the rates of return expected by potential innovators. For nations that do not adequately address such market failures, development and subsequent commercialization of technologies are slowed. A number of these barriers occur because several elements of the typical industrial technology have the characteristics of infrastructure. This infrastructure facilitates—and, in fact, makes possible—the development, production, and commercialization of new technologies.

THE ROLES AND IMPORTANCE OF TECHNOLOGICAL INFRASTRUCTURE

Infrastructure has, for centuries, been recognized as essential to sustained economic growth. Its form has changed, as have the economic assets it supports. Today, the complexity and technological content of infrastructure have grown along with the evolution of technologies that form the basis for products, production processes, and services.

Public infrastructure investment results in considerable economic impacts. Huge investments in infrastructure during the last half of the ninteenth century, much of it funded by foreign sources, accelerated the industrial revolution in the United States and thereby greatly abetted that country's ascendancy to the status of the world's leading industrial economy for most of the current century. This investment in so-called traditional infrastructure—roads, bridges, canals, railroads, and communications—aggregated heretofore regional markets into a national one. The ability to sell into one large national market as well as to cost effectively access inputs from other geographic regions increased the scope as well as the scale efficiency of production.

In the last several years, economists have begun to systematically study the roles and economic impacts of infrastructure. Analyzing the substantial drop in productivity growth in the U.S. economy between the period 1950–1970 and the following fifteen year period (1970–1985), David Aschauer, (1988, 1989) found that "fully 1.0 percentage points of the total decline in (the average annual growth rate of) productivity of 1.2 percentage points can be attributed to the neglect of infrastructure."[1]

Other economic studies (Aschauer, 1989; Erenburg, 1993) have shown that government spending on economic infrastructure in general has significantly leveraged private investment. Erenburg (1994) analyzed data on long-term trends in private and public investment and the effects of public infrastructure investment on productivity and output growth rates. Significant findings are:

- Public investment in infrastructure has fallen below its historical average in the last several decades.
- Econometric analyses estimate that maintaining the historical average rate of growth in government spending for infrastructure would have added 4.0 to 6.0 percentage points to the growth rate of private equipment investment.
- This additional investment would have added 1.3 to 1.9 percentage points to GDP growth above actual rates.
- Erenburg (1994, p. 21) also estimated that due to the slowdown in public investment, the capital stock of the United States in 1990 was $662 billion below its historical average when comparing the public capital stock to private capital stock. In terms of annual rates of investment in this public capital stock, the United States was $120 billion below its historic ratio of public investment spending to the private capital stock.

Given the leveraging effect of public investment on private investment and the subsequent increase in secular growth of productivity and output, the trends in the two categories of investment, as shown in Figure 1.1, should be cause for concern.

Several studies have been undertaken in recent years with respect to the impacts of government R&D spending on private-sector R&D investment. Leyden and Link (1992) modeled the process by which industry-funded and government-funded R&D interact within the corporate laboratory. From survey data, they estimated that a $10 million dollar increase in government R&D would result in a $27.5 million increase in private R&D spending. They also concluded from their survey that R&D personnel in firms receiving

Figure 1.1
Public-to-Private Captial Stock Ratio
Public to Private Capital Stocks, 1947–1991 in $1987

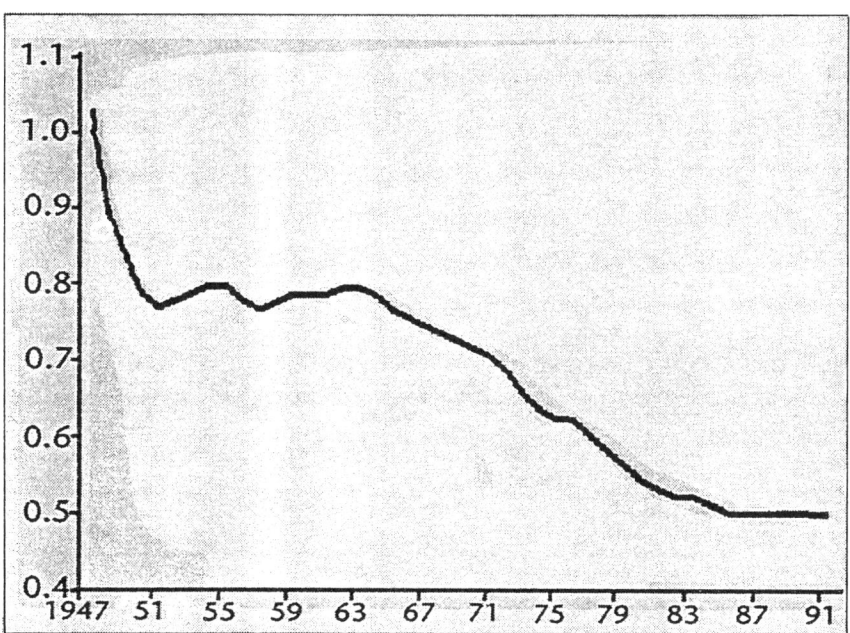

government R&D funds tended to allocate on average 15 percent more of their time to "sharing of technical knowledge."[2]

Mamuneas and Nadiri (1993) developed an econometric model of the interactive effects of public and private R&D expenditures in the manufacturing sector. They found that "the effects of publicly financed R&D are overall significant and vary across industries." Specifically, their results show that publicly financed R&D increases efficiency in terms of unit cost savings, thereby increasing an industry's competitiveness. This effect occurs even though the publicly financed R&D substitutes to an extent for privately financed R&D, at least in low R&D-intensive industries. In other words, even though the availability of public investment substitutes for some private investment, an overall efficiency gain occurs.

Thus, these two studies support the existence of a complementary relationship between public and private R&D investment, as was found for investment in general. Moreover, much of U.S. public R&D funding has been only tangentially directed toward support of industry's technology commercialization objectives. When public R&D funding is designed to directly support market strategies, the resulting infrastructure has even greater economic leverage.

Such market-oriented, technology-based infrastructure investment by government not only makes private domestic investment more productive, but it also increases the efficiency by which foreign technology is absorbed into the domestic economy. It is not a coincidence that the Coe, Helpman, and Hoffmaister (1995) study found that Singapore's productivity was boosted by foreign technology capital to a much greater extent (0.22 percent) than the average for 77 developing economies (0.04 percent). The Singapore government has invested heavily in advanced infrastructure, including information infrastructure.[3] An increasing number of nations are making substantial investments in market-oriented technological infrastructure. In fact, this category of public investment is becoming more important as a national competitive strategy, not only because it is necessary for technology-based economic growth to take place, but also because it is a relatively immobile economic asset and thereby induces a net inward flow of investment for the domestic economy.

Old Model of Government Support for Technology

For most of the postwar period, the economic growth policies of industrialized nations were based on a simplistic and inaccurate view of technology. The typical industrial technology was viewed as being homogeneous in nature—a "black box," so to speak.

Within this framework, basic science is a "public good" and its provision is therefore primarily the responsibility of government. This principle has been accepted since the Vannevar Bush report, Science the Endless Frontier, published shortly after the end of World War II. That report led to the creation of the National Science Foundation a few years later in 1951, whose mission is to support basic research.

However, technology, when developed as an asset for economic activity, was deemed to be a purely private good. Government therefore was not thought to have a role in its development or diffusion throughout the economy. A major reason for adherence to this model was the fact that a rush of new technologies appeared in the postwar period and the U.S. economy was a prime mover in commercializing most of them. Because national defense programs funded considerable technology development, which eventually "spilled over" into private markets to varying degrees, the efficiency of infrastructure support for technology development and commercialization was not an issue. In fact, the emergence of new technologies, in part from government R&D, led to the belief that technologies could be "pushed" into the marketplace, rather than "pulled" by actual demand.

The New Technology Infrastructure Model

Over the past two decades, the realities of global competition have led to the evolution of a conceptual model of technology-based economic activity that recognizes the fact that the typical industrial technology consists of a number of elements—some private, some public, and some a mixture. The "public" or "collective consumption" of several of these elements gives them an infrastructure character. As the typical product and service have become increasingly dependent on the rapid and effective incorporation of technology, industrialized nations are making substantial investments in new technology-based infrastructure. In fact, it is coming to be viewed as an essential element of a competitive economy.

Technology infrastructure is an element of an industry's technology that is jointly used by competing firms (Tassey, 1991, 1992; and Justman, 1995). One category of this advanced infrastructure is what has come to be called an industry's generic technology. Generic technologies are the core product and process concepts from which specific commercial applications are developed through subsequent applied R&D.

Typically, a generic technology embodies a laboratory-proven concept, but not the subsequent market-specific products and processes that are eventually derived from it. Achieving a level of generic technical knowledge sufficiently reduces technical risk in most cases to allow investment decisions—positive or negative—to be made with respect to subsequent applied R&D. The substantial technical risk, coupled with large economies of scale or scope that frequently characterize generic technology research, lead to substantial market failure at the company level and thus to underinvestment.

Thus, much generic technology has come to be viewed as a public or quasi-public good (in effect, as infrastructure) and is therefore developed jointly by firms that compete against each other later in the technology life cycle. The rationale is that sharing the early-phase R&D results (the generic technology) and having access to it earlier in the competitive life cycle is preferable to an "all-or-nothing" strategy in which each firm tries to independently develop the generic technology as a totally proprietary asset. Failure to see the infrastructure character of generic technology often results in cost and time disadvantages with respect to foreign competition, in turn resulting in market growth opportunities being permanently lost.

A second category of technology infrastructure includes the various techniques, methods, procedures, etc., that are necessary to implement the firm's product and process strategies. Methods such as total quality management (TQM) can be differentiated upon implementation within a firm, but they must be traceable back to a set of generic underlying principles, if customers are to accept claims of product quality. For example, demonstration to customers of compliance with ISO 9000, a set of international quality assurance standards, must be derived from a set of common principles even though implementation within the firm can be customized to fit internal production strategies. Thus, the ISO 9000 standards constitute a form of infrastructure. And, especially for small firms in traditional industries, business and technology assessments—either provided directly or through referrals to private consultants—can mean survival.

A third type of technology-based infrastructure consists of a set of "technical tools" for making the entire economic process more efficient, or, in some cases, possible in the first place. Collectively, these tools are called infratechnologies (Tassey, 1991, 1992). For most industries, all phases of R&D, production, and market development are supported by a range of infratechnologies.

As indicated in Figure 1.2, infratechnologies are ubiquitous in terms of their scope of impacts on the typical technology-based industry. This element of technology infrastructure becomes embodied in or supports generic technology and its applications. Infratechnologies also provide the technical basis for a number of types of standards, which directly affect process and quality control at the production stage. Moreover, similar categories of standards are typically essential for the efficiency of market transactions through the reduction of performance risk to the buyer of advanced products and services.

Infratechnologies fall into four general categories:

- Scientific and engineering data that are used for conducting R&D, controlling production, and consummating market transactions (the latter applications are especially important in continuous process technologies, such as chemical production and petroleum refining).
- Measurement and test methods that are essential to conduct state-of-the-art R&D, monitor production, and execute market transactions (product acceptance testing).
- Production practices and techniques, such as process models for understanding relationships among production parameters, that thereby allow more efficient design and control of production processes.
- Interfaces that permit the efficient physical and functional combinations of components into manufacturing and service systems.

As the complexity and pervasiveness of technology grow, infratechnologies are becoming ubiquitous in the modern economic system. As the arrows in Figure 1.2 indicate, this category of infrastructure is embedded in each of the three major stages of technology-based economic activity (R&D, production, and marketing).

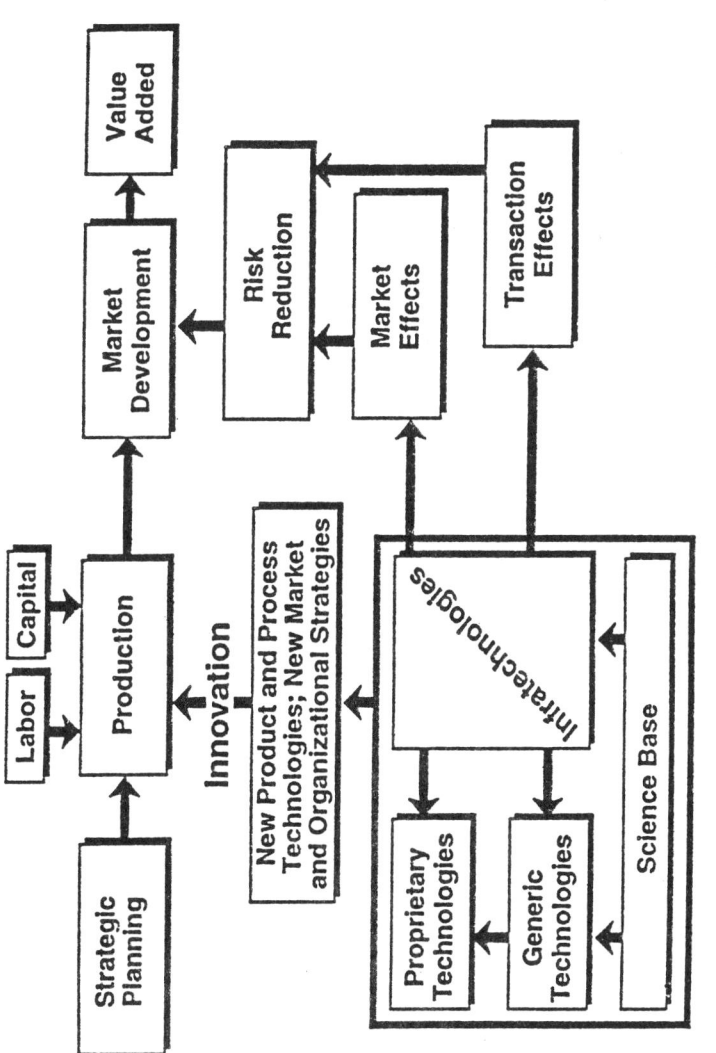

Figure 1.2
Disaggregated Technology Growth Model

PUBLIC-SECTOR INVESTMENT IN TECHNOLOGICAL INFRASTRUCTURE

As described earlier, an industry's technology has a "base" or generic component, which individual companies draw upon to develop specific market applications. Consequently, corporations that compete against each other can nevertheless "share" the same generic technology.

Furthermore, the increasing complexity of the typical technology, coupled with the greater demands from customers for quality, reliability, and shorter response times, have combined to create the need for a set of "technical tools" or infratechnologies. As described in the previous section, these infratechnologies include databases of the properties of advanced materials, measurement, and test methods for controlling production and certifying performance before customer acceptance, and the technical basis for defining the interfaces among the many components of complex systems that dominate the automated factory, information technologies, and so on.[4]

For a number of reasons, some of which have been cited above, individual firms and even entire industries will underinvest in these infrastructure elements. The fact that spillovers occur in technology-based industries has led some analysts to focus on the size of the "gap" between the innovator (private) and spillover (social) rates of return (RoR) as an indicator of the magnitude of market failure and subsequent underinvestment, and hence of the desirability of government intervention in that particular market.

However, a more accurate characterization of the rationale for a government role is given by the following two conditions: (1) if the spillover RoR is high, the technology is important to the economy, and (2) if the innovator RoR is low (relative to the hurdle rate), underinvestment will likely occur. Thus, the innovator RoR determines whether or not underinvestment occurs and the gap between it and the spillover RoR determines how important to the economy underinvestment might be, if it is occurring. But, the gap by itself does not necessarily mean that underinvestment and hence a market failure exists.

Another problem with the literature on market failure is that it confuses ex ante and ex post rates of return. The former refers to the expected RoR that a firm demands for any one project, given its assessment of risk and payoff. The later refers to the actual or realized RoR. Often, the high rates of return from investments in technology are cited as necessary to compensate investors for the relatively high risk associated with this class of investment. However, the RoR cited in this literature are ex post returns. That is, they have already occurred, so that the risk associated with the R&D investment is already reflected in the reported RoR. Thus, knowledge of these realized RoR should provide sufficient incentive to generate optimal or at least higher levels of investment in R&D.[5]

So, the question is why, knowing that the ex post average rate of return is higher than for alternative investments, do firms not invest more in R&D. After all, in a perfectly functioning capital market, funds should flow to the higher-return investment areas until the ex post return is bid down to the average for all investments.

The answer is that the rate of return expected by the innovating firm determines investment decisions by individual firms. When this innovator RoR is not high enough, compared with less risky alternatives—namely investment in plant and equipment, embodying existing technology—the latter category of investment is often selected. These relatively low rates of return to private investment in technology are the result of serious and persistent market failure—and not just of the spillover type. The categories of market failure that regularly impact the flows of investment funds into technology development are shown in Table 1.1.

These market failures either prevent objective risk assessment or lower the expected return relative to risk. Furthermore, market failures seem to be occurring at two levels: (1) the overall level of R&D investment, and (2) within specific categories of R&D.[6]

Such barriers occur when (1) technical risk is high, (2) the capital intensity of the research process is substantial, (3) the time to completion of the R&D, and hence the time to commercialization and the realization of a cash flow from the R&D investment is long, (4) the scope of potential markets is broader than the scope of existing market strategies, so that individual firms do not project economic benefits from all potential market applications of the technology, (5) the technology involves a significant element to which the assignment of intellectual property rights is difficult, (6) industry structure raises the cost of market entry for applications of the technology, and (7) the complexity of the technology makes agreement with respect to product performance between buyer and seller costly.

The existence of risk has been used as the rationale for government support of R&D. However, this is an oversimplification of the realities of technology-based competition and has therefore contributed significantly to the lack of consensus with respect to government roles in support of this sector of the economy. In fact, risk derives from a number of sources and can vary substantially in magnitude.

Some risk is simply inherent in technology-based markets and the rewards of successful innovation can and do compensate firms for assuming reasonable levels of risk. However, technical risk, especially in the early phases of development can become prohibitive because of the intrinsic nature of the technology being pursued. Such technical risk can arise from either complexity

Table 1.1
Examples of Technology-Based Market Failures and Government Policy Responses
(With Approximate National Government Contribution to Annual Budget)

Market Failure -- infrastructure element:	High Risk, Long Time to Market, Scope of Markets -- generic technology	Non-Capturability, Industry Structure, Transaction Costs -- infratechnology	Technology/Market Strategy Mismatch -- applied R&D	Technology/Market Strategy Mismatch -- technology transfer
United States	NIST: Advanced Technology Program (50% cost sharing with single firms and consortia) -- $340M	NIST: laboratory research (some cost-shared consortia) -- $260M	none	NIST: Manufacturing Extension Partnership (MEP) -- $75M of $150M total
Japan	MITI: Industrial Science and Technology Frontier Program -- $260M; Bureau programs (RWCS, IMS, FGC, etc.) -- $90M;	MITI: laboratory research (some collaborative research) through AIST's 15 laboratories -- $500M	MITI/MPT: Key Technology Center (70% cost-sharing of joint ventures) -- $250M	regional applied R&D/technology transfer centers (Kosetsushi) -- $300M of $1,300M total
European Community	EC: Framework IV Program -- $3,000M (50% cost sharing), with approx. $1,500M for "enabling" technologies; individual country programs; e.g.: Germany -- $1,300M	individual country laboratories; e.g.: Germany -- $180M, U.K. -- $65M	multilateral: EUREKA (35% cost sharing) -- $1,200M	individual country programs; e.g.: Germany (Fraunhofer Institutes) -- $200M of $650M total

or from research requirements that require multidisciplinary research teams not generally available within individual firms.

High technical risk can also translate into high commercial or market risk when the required R&D is capital intensive, so that the minimum cost of conducting the research is high relative to a firm's overall R&D budget.

Time also translates into high levels of risk. Private firms apply a relatively high discount rate to all investment projects, meaning that the benefits have to occur within a few years or they are frequently not valued high enough to result in an acceptable projected RoR.

Unacceptable risk is also created when certain elements of an industry's technology must be shared among firms in that industry in order to have high economic impact (a standardized test method, for example). The nonproprietary character of such technology elements leads to underinvestment because the benefits cannot be captured by a single firm.

Finally, the broader market environment in which the technology will be sold can significantly reduce incentives to invest in its development and commercialization. Most technology-based products in today's complex world are part of larger systems of products (examples are an automated factory and a communication network). If a high-tech firm is contemplating investing in the development of a new product but perceives a risk that the product—even if successful—will not "fit" or "interface" with other products in the system, the additional cost of attaining compatibility may reduce the expected RoR to the point that the project is not undertaken.

When the above set of factors leading to underinvestment in technology-based industries is compared to simplistic attempts to rationalize government roles—namely, "high risk"—it is easy to understand the prolonged and inconclusive debate that has persisted in the United States.

INFRASTRUCTURE STRATEGIES IN OTHER INDUSTRIALIZED NATIONS[7]

For approximately the past fifteen years, most industrialized nations have been debating the respective roles of industry and government in supplying their domestic industries with the infrastructure necessary to support technology-based economic growth. Over this period of time, these nations have moved toward a consensus framework represented by the conceptual model in Table 1.1. In fact, the United States is the only nation currently still debating the rationales for these major infrastructure roles.

Programs can be found in most nations that target each of the major elements of infrastructure. Where an economy is too small to effectively implement technological infrastructure strategies, it joins with neighboring countries and provides the infrastructure on a regional basis (such as the European Community's Framework Program and multinational EUREKA

Program). The content of individual programs differ, that is, the individual policy mechanisms used by different nations or groups of nations vary, but each nation is addressing the major categories of market failure.

When the market failure is general risk aversion, so that aggregate investment in R&D is inadequate, tax incentives are the usual policy response because this mechanism can be reasonably efficient in reducing the cost of overall R&D.

Since 1981, Congress has provided a tax credit for industry-funded "research and experimentation" (R&E). However, its coverage is very similar to the National Science Foundation definition of "research and development" (R&D). The R&E tax credit is incremental in structure. It is calculated as 20 percent of the increase in R&D over a "base" R&D number, where the base is calculated by multiplying the firm's average R&D-to-sales ratio for the years 1984–1988 by the average sales of the four years prior to the year for which the credit is being calculated.[8] Because it is a credit, as opposed to a deduction, it applies directly dollar for dollar to taxes owed. The important point for policy analysis is that it tends to leverage investment in R&D generally and therefore implies a market failure in the category of general risk aversion.

Other industrialized nations provide similar and additional tax incentives. Japan, for example, also has an incremental 20 percent tax credit for R&D expenses exceeding the highest previous year's expenditure (up to a maximum of 10 percent of total corporate tax). However, Japan also allows small and medium firms to deduct 6 percent of all R&D expenses up to a maximum of 15 percent of total corporate tax. Furthermore, a 6 percent tax credit is offered for R&D expenses incurred in joint projects with national research institutes. To further promote cooperative R&D, Japanese firms benefit from accelerated (one-year) depreciation of equipment used in joint R&D projects.

Firms tend to either reclassify existing R&D to take advantage of the tax incentive or simply ignore it. Reclassification of R&D is an example of a "leakage" effect associated with an ineffective policy mechanism; that is, costs (tax expenditures, in this case) are incurred without achieving the desired effects on private-sector behavior. With respect to rejecting a tax incentive, this has been the case with the tax credit available for corporate funding of university research. It has been little used and the reason seems to be that to the extent companies want to conduct longer-term, generic technology research, they do it internally or through a consortium of firms and government laboratories.

In general, tax incentives are good at increasing the type of R&D that industry is already doing, but not so effective at restructuring the composition of R&D. Realizing this, most industrialized nations use several types of policy mechanisms, only one of which is tax based.

Australia, for example, has used a strong tax incentive consisting of a 150 percent deduction from taxable corporate income for business expenditures on

R&D. At current tax rates, the cost of R&D to Australian is cut in half. The result has been a dramatic doubling of corporate R&D in the past decade. However, even in this rather extreme case of the use of tax policy to increase aggregate R&D spending by industry, the Australian government has realized that the composition of R&D is a separate problem and has implemented programs to remove this latter market failure (OECD, 1994, pp. 39–43).

Thus, government policy in industrialized nations has evolved various types of direct R&D funding programs to address these more focused market failures. In the United States, Congress created the Advanced Technology Program (ATP) in 1988 as one element of the National Institute of Standards and Technology (NIST). Its purpose is to help industry remove market failures that commonly appear in the early phases of industrial R&D.[9] The program does this by soliciting project proposals from industry, conducting both technical and business evaluations of the proposals, and then cost-sharing the funding of the selected projects. The ATP was initially funded at a very low level. However, its budget was rapidly escalated to $430 million in 1995, before being reduced to $340 million in that same year as part of a budget recision bill.

Japan

In Japan, the primary, but not the only, agency funding early-phase generic technology research is the Ministry of International Trade and Industry (MITI), which is generally credited with bringing Japanese technological capability up to global standards. MITI funds projects through several mechanisms.[10] One of these mechanisms is a series of ongoing project areas in which generic technology research is funded. The best known of these are the Large-Scale Project and the Next Generation Project. A third somewhat smaller program is the Medical and Welfare Equipment Technology Project.

Beginning in the 1960s, the Large-Scale Project has subsidized approximately thirty major projects over several decades with a collective expenditure of approximately $2.5 billion. These projects range from large-scale integrated circuits to high-performance electric car batteries to desalination equipment.

The Next Generation Project was initiated in the early 1980s to fund "enabling" technologies for key industries of the future, such as new materials, biotechnology, new electronic devices, superconductivity, and new generations of computer software.

The major project areas under each of these headings were reorganized in 1993 under an umbrella program called the Industrial Science and Technology Frontier Program. This program is funded at about $260 million per year and focuses on subsidizing the early phase development of enabling industrial technologies. It is thus similar in objectives to NIST's Advanced Technology Program (ATP).

MITI also manages Japan's major energy and environmental technology programs such as the Sunshine and Moonlight projects (begun in the 1970s) in which a total of $4.5 billion has been invested to advance solar, geothermal, coal, and hydrogen energy technologies, as well as energy conservation technologies, and the $150 million Global Environmental Industrial Technology Project to develop technologies to reverse ozone depletion and reduce greenhouse gas emissions. These programs were also reorganized under an umbrella program—The New Sunshine Project. Funded at about $550 million per year, it supports advanced energy and environmental technology research programs. These programs are managed for MITI by one of Japan's numerous "public" or "quasi-government" corporations—the New Energy and Industrial Development Organization (NEDO).

In addition, MITI occasionally sponsors R&D programs through its bureaus. For example, in 1981, the Machinery and Information Industries Bureau launched the ten-year, $400 million Fifth Generation Computer Project. The bureau currently sponsors two major "international" R&D programs: the 10-year, $500 million Real World Computer System (RWCS) Project to develop information processing systems capable of humanlike analyses and the ten-year, $1 billion Intelligent Manufacturing System (IMS) Project to increase software and computer integration capabilities for next-generation manufacturing technologies.

A number of MITI projects (such as the VLSI Project) have been successful in upgrading Japanese industrial technology capability to world-class levels. In cases where the projects have attempted to set a standard, such as in the general area of information technology, they have been less successful. With respect to the latter, the Sigma Project (1985–1990, $150 million) attempted to set a new standard for computer applications software, TRON (1984–1994) attempted to develop a new generation of microprocessors, the Fifth Generation Computer Project (1982–1992, $250 million) attempted to develop new computer architectures based on artificial intelligence concepts, and the ongoing Real World Computer System Project (1992–2002, $500 million) is attempting to develop new computer hardware and software architectures.[11] In general, these efforts have been viewed as relatively unsuccessful by a number of analysts, both inside and outside Japan.

To the extent that generalizations can be made from the Japanese experience, one question to be asked is why have government-supported projects that are designed to catch up with foreign competition in general been fairly successful, while projects that are pursuing next generation technologies, and hence technological innovation (first-to-market strategies), have not worked particularly well.

One possible answer is that the Japanese have never quite grasped the model represented in Table 1.1. When catching up is the objective, issues such as the phase in of the R&D process, being subsidized by government funds and assignment of intellectual property rights, is not as important. The Japanese VLSI project in the late 1970s was an attempt to pull Japanese semiconductor technology up to the level of U.S. firms.[12] Although one hundred or so patents resulted from an expenditure by MITI of $300 million over four years, and a few of these patents turned out to be commercially significant, the major impact was a general upgrading of industry's technological capabilities. This occurred primarily through greater commitment to semiconductor R&D by individual Japanese electronics firms. In fact, little cooperation among participating firms seems to have occurred during this project.

In such situations, the focus on early-phase or generic technology was not emphasized. Thus, the market failures that occur at this phase of R&D were not a strong policy focus, as is the case when a breakthrough to a new generation of technology is being attempted. Similarly, the issue of intellectual property rights was not particularly important as the technological capability being pursued was already generally available elsewhere in the world.

On the other hand, the information technology projects are attempts to establish technological leadership for Japanese industry. By the time these projects were initiated, Japanese firms had achieved world-class status with respect to technological capability. As a result, they are more concerned with their established, as well as their future, competitive positions. Currently, the typical Japanese electronics firm seems willing to participate in a government-supported project at the generic technology phase of R&D. But, they are less inclined to cooperate beyond the point (i.e., the applied R&D phase) at which they begin to pursue proprietary market applications that will compete against other Japanese firms.

In both cases, Japanese firms are now more conscious of intellectual property that arises from R&D. Even though generic technology research is a distance from the eventual innovations based on it, intellectual property is produced that can be owned by at least several firms collectively. Unlike NIST's Advanced Technology Program (ATP) and the European Community (EC)'s Framework Program, which assign property rights entirely to project participants, the Japanese government has been slow to recognize this requirement for industry participation. Until recently, the Japanese government believed (as did the U.S. government until the mid-1980s) that government-funded research should be treated as a "pure" public good; that is, the results of the research should be made available to all domestic firms interested in the technology.

Beginning in 1991, the Japanese government finally began to address the issue of insufficient incentives for participation in research consortia. It adopted a partial patent-sharing mechanism by which participants (including foreign firms) in MITI and STA (Science and Technology Agency) research projects can now share up to 50 percent of resulting patents. Each participant's share is determined by its technical contribution to the research leading to the patents. This change has brought Japanese government-sponsored research programs closer to the U.S. (ATP) or European (Framework, EUREKA) models. However, it still does not completely close the incentive gap.

In addition to the motivation of increasing Japanese companies' participation in government-sponsored research projects, the Japanese government has also made an effort to increase foreign participation in certain designated international research projects. For example, as part of NEDO's implementation of MITI's Industrial Science and Technology Frontier Program, it has been gradually implementing the 50 percent patent sharing policy (which applies to foreign as well as domestic firms).

Eighteen foreign companies, including ten U.S. companies, currently participate in ten Frontier projects. However, restrictive factors still limit foreign participation in NEDO and other Japanese research projects. Japan's patent provisions still include restrictive government rights, such as the right of refusal for technology licenses, the right to collect royalties from project participants, and the right to set preferential royalty rates and determine those who can get those rates. Moreover, U.S. firms continue to have difficulty gaining access to critical information during the planning stage for some research projects, making it harder to gain entry.

MITI also operates fifteen laboratories through its Agency for Industrial Science and Technology (AIST) with an annual budget of about $500 million. These laboratories fund or conduct both generic technology and infratechnology research. A significant portion of the research is conducted on a cost-sharing basis with private companies. University researchers also participate in projects.

A lesser known but important mechanism by which Japan funds industrial R&D is the so-called public corporation (PCs). There are ninety-two of these entities engaged in promoting various industries. Of this number, twenty-four implement some aspect of S&T policy as their primary or one of their primary missions. Within this latter group, nine conduct or support research, eight finance technology development, three manage government gambling operations but provide funds (from gambling revenues) for R&D promotion, two collect and disseminate S&T information, and two manage "development assistance" to promote Japanese technology-based exports.

Except for those PCs that generate a portion of their funding activity from internal operations (e.g., the gambling management PCs), the main sources of funds are the Fiscal Investment and Loan Program (FILP), which draws funds from the Postal Savings and Pension programs, and from the General Account Budget.

Of the twenty-four S&T-related PCs, eighteen received either low-interest loans from FILP or direct funding from the General Account Budget. In 1993, FILP provided three quarters of the combined budget of about $130 billion for these eighteen PCs and only 9 percent came from the general budget, with the remainder coming from retained earnings and private sources. Because FILP funds have grown steadily, these eighteen PCs have been able to increase spending, in spite of tight government budgets that have adversely affected other programs.

Several of the better known PCs include Nippon Telephone and Telegraph (NTT), NEDO, and the Research and Development Corporation of Japan (JRDC). NTT is an example of the Japanese government divesting a portion of its ownership in a PC's asset structure.

As part of a changing technological infrastructure strategy, MITI is moving away from its emphasis of the past several decades. This strategy funded applied R&D in a limited number of large high-tech firms, where the research was conducted primarily within individual corporate laboratories. The trend now is to emphasize early-phase generic technology research conducted cooperatively, much of it in AIST laboratories with participation of industry, university, and government researchers.

At the same time, MITI is decentralizing the conduct of generic technology research by increasing support for regional R&D. This is being accomplished through seven regional government industrial research institutes (GIRIs). MITI is pressuring the GIRIs to take a leadership role in developing regional superiority in one or, at most, several manufacturing technologies (examples are specific advanced materials such as ceramics, biotechnology, and optoelectronics). MITI wants each GIRI to become the R&D "core" of a regional manufacturing base and is providing them with R&D funding and a package of tax and other financial incentives to induce each regional center and surrounding small and medium enterprises (SMEs) to specialize in the chosen technology. MITI is also promoting closer collaboration between a GIRI and the manufacturing technology transfer centers (Kohsetsushi) in that region. In fact, the bulk of Tokyo's R&D budget provided to local governments—about $300 million—is funneled into Kohsetsushi to promote regional R&D investment.

Finally, in 1985 Japan established an interministerial R&D program called the Key Technology Center (KTC).[13] Under the joint management of MITI and the Ministry of Posts and Telecommunications (MPT), the KTC was capitalized with dividends that the Japanese government earns from the NTT stock that it retains. It has an annual budget of approximately $250 million, of which about 75 percent goes to a "capital budget" and the remaining 25 percent to a loan program.

The majority of the capital budget is used to fund joint ventures (JVs) through equity investments. An equity investment by the KTC in a JV can provide up to 70 percent of project operating funds for up to seven years. Annual funding is contingent upon "acceptable" technical progress. The KTC has the right to recoup its investment either through a sale of its equity in the JV or through a share of the JV's earnings, once the technology is commercialized.

Since its inception, the KTC has invested in fifty-eight R&D joint ventures in areas such as new materials, biotechnology, automated manufacturing equipment, electronics, communications signal processing, network design, etc. Each JV applying for funds must submit both a research plan and a business commercialization plan (as is the case for applicants to NIST's Advanced Technology Plan). Stated market failures targeted by the KTC are high technical risk, market applications requiring "fusion" of technologies from distinctly different industries (market-technology mismatch), scale-up problems (high capital intensity), and missing complementary infratechnologies (standards).

The loan program can provide up to 70 percent of total R&D costs (operating plus facilities) to individual companies for up to five years. In the latter case, if the project is successful, the loan plus interest must be paid back; if unsuccessful, only the principle is recouped.

Unlike other Japanese government R&D programs, intellectual property rights are assigned completely to participating private companies in both the capital budget and loan programs. However, the incentive to participate may nevertheless be reduced by the equity interest taken by the government for funding under the capital budget.

Europe

Individual European countries have had technology infrastructure programs of various types for decades. However, most of these economies do not have sufficiently large internal markets to capture enough of the spillovers from typical technology infrastructure research programs to rationalize government investments that have the minimal efficient scale or scope. Consequently, the European Community (EC) has invested in a series of "Framework" programs to collectively develop needed technology infrastructure.[14]

Earlier framework programs in the 1980s suffered from a poor understanding of market failures that occur in technology-based markets. In particular, managers of these programs did not clearly articulate that the targeted early-phase generic technology research should not be expected to directly yield commercial products. They were therefore unable to adequately respond to criticisms of insufficient commercial success.

By the time Framework IV was implemented in the early 1990s, a conceptual model similar to that in Table 1.1 was more firmly in place. This has not only increased general support for the program but has led to more effective project selection. It also allows this program to be contrasted with the EUREKA program, which is operated independently of the EC by the European governments.

EC documentation describes the objective of Framework IV as providing financial support for "precompetitive advanced technologies with high potential for commercialization." Unlike early MITI programs, small and medium firms are encouraged to participate. EC managers set research area priorities with significant input from the scientific and business communities (similar to the ATP approach to establishing "focus" technology areas). Individual research proposals come from industry and are selected through a competitive process (as is done in ATP).

More than half of the annual budget of $3 billion is allocated to supporting the development of "enabling" technologies. This implies that economies of scope is at least one of the market failures being targeted. That is, the full range of market opportunities offered by the new technology is not currently within the strategic foci of individual firms, thereby contributing to underinvestment in the generic technology.

The Framework IV Program only funds consortia, which can include universities, and private and public research institutes in addition to private firms, and must include participants from more than one nation. Foreign participation requires an "integrated presence" within the EC economy. Eighty percent of the program's funds are awarded on a cost-sharing basis, with private firms and any participating national laboratory contributing at least 50 percent of the total project funds. Intellectual property is owned and shared by the consortium members.

The largest single program under Framework IV is ESPRIT (European Strategic Program for R&D in Information Technology), which has funded over six hundred projects with an annual budget of approximately $500 million. ESPRIT's objective is to support "precompetitive research in information and communication technologies." The scope of targeted technology areas includes microelectronics, information processing systems and software, and manufacturing integration. Assistance in standards development is a cross-cutting secondary objective.

ESPRIT is a good example of how the conceptual model of technology infrastructure has evolved in Europe. In recent years, user industries (those industries whose firms purchase the products or services based on the new technology) have appeared as consortium members along with firms in the industry that are the eventual supplier of the technology. Such promotion of "virtual" vertical integration not only increases the effectiveness of research project design, but also addresses industry structure problems that are present in many European industries. ESPRIT has also increased funding of clusters of projects to achieve both scope and depth in a particular technology.

Finally, European countries operate a multicountry collaborative R&D program independent of the EC called EUREKA.[15] Because of its lack of central management, no strategic planning takes place (for example, no clusters and no competitive selection of projects within clusters). There appear to be approximately five hundred ongoing projects with a total annual funding of $1.2 billion. EUREKA projects are somewhat more toward the applied R&D phase than the Framework programs and thus are arguably complementary to the Framework projects. As a result of the greater market orientation of these projects, national governments contribute a maximum of only 35 percent of the total project cost. The largest single program area under EUREKA is JESSI (Joint European Submicron Silicon Initiative), which is seeking to develop a sixty-four megabit memory device by 1996. Funding for this project is 50 percent industry, 25 percent national governments (largely France and Germany), and 25 percent EC.[16]

SUMMARY

In addition to the above technological infrastructure programs of established industrialized nations with whom the United States competes, newly industrialized countries are rapidly setting up similar infrastructure capabilities. Korea, Taiwan, China, and Singapore are examples. Thus, the industrialized world continues to refine the "mixed economy" model of economy growth, based in the model represented in Table 1.1. In such a framework, the typical industrial technology has several discrete elements, some which have distinct public-good attributes.

Ironically, this global trend is occurring at the same time that the United States is debating whether or not to significantly curtail its nascent technological infrastructure capability.

The programs in place in Europe and Japan to address various imperfections in technology-based markets are summarized in Table 1.1. The approximate current annual funding for these programs is given for the national government. Where appropriate, the percentage of total funding provided by government for industry-government, cost-shared projects (typically generic technology and applied R&D phases) is provided. Where significant regional or local government funds are involved (typically for technology transfer programs), the national government contribution and the total funding are given.[17]

As indicated in the table, the United States is the only nation that does not have an ongoing program that explicitly supports its domestic industry's investment in specific applied R&D projects, although it has supported ad hoc projects such as SEMATECH and the current Partnership for a New Generation Vehicle (PNGV) that include applied research.

In summary, all nations that hope to compete in the global economy must have substantial technological capability. Because of the complexity of the typical industrial technology, this capability will have to come from a combination of industry and government sources. Every industrialized nation attempts to implement this "mixed economy" model. The specific methods of implementation vary and so apparently do the outcomes as measured by shares of world markets and rates of economic growth. To this end, government's role is twofold: (1) to provide a macroeconomic environment that is conducive to high rates of private-sector investment that includes sufficient risk taking, and (2) to provide infrastructure support at more microeconomic levels in order to efficiently leverage private-sector investment in specific technological areas.

NOTES

1. Use of the term "infrastructure" continues to suffer from two distinctly different definitions. One version characterizes infrastructure very broadly, including all industries that supply the industry in question. Thus "supplier industries" or "a network of firms that produce goods and services" are frequently used to define the infrastructure for a particular industry. Unfortunately, this makes virtually all industries infrastructure to other industries and hence confuses analysis of true infrastructure roles.

A more reasonable and functional definition is to define infrastructure, not in terms of industries, but as products and services that are consumed more or less equally by competing economic entities (firms or individuals). According to this latter definition, a road used by competing trucking companies is infrastructure as is a communications network used by competing financial service firms, but industries that supply materials or labor to build roads or equipment and software to construct a network are not infrastructure (the latter supply competing private goods or services that are used to create infrastructure).

2. "Sharing" refers to the dissemination of technical knowledge through publishing articles, presenting papers at conferences, etc.

3. Government R&D is viewed by much of the economic literature as complementing private R&D, but the rate of return has frequently been estimated to be significantly lower than the average for private-sector R&D (Nadiri, 1993). This finding is not surprising, given that these studies focused on U.S. government R&D, which historically has been largely mission-oriented (defense, space, etc.). Although spinoffs to the private economy from mission-oriented government R&D do occur, it is generally recognized that such R&D is not the most efficient way to complement private R&D for pursuit of economic growth objectives. When market failures affecting private investment in R&D are directly targeted, the rate of return to government R&D should be much higher. That this is the case is strongly indicated by the data reported later in this report.

4. The recognition of the role and importance of infratechnologies for competitiveness is identified in both the 1994 Economic Report of the President (p. 191) and in Science and Technology: A Report of the President, 1995 (p. 17).

5. According to basic economic theory, additional resources should be allocated to R&D projects with progressively lower expected RoR (the next project selected is the one with the next highest expected RoR) until the ex post RoR has fallen to the hurdle rate (also called the opportunity cost of capital and is approximately the average return on investment generally).

6. Because most of industry's R&D expenditures are for applied research and development, it is primarily the returns on investment in these later phases of the R&D process that are reflected in the cited economic RoR studies. The economic studies that have been done of certain categories of technology infrastructure (such as the infratechnologies produced by NIST) show higher rates of return, indicating the economic importance of removing the market failures. However, these latter estimates have been derived from case studies. Thus, they cannot yet be considered to represent average returns from infratechnology investment until a sufficient number have been done to cover significant portions of the total infratechnology investment for a representative number of technologies.

7. The material in this section has been collected from a variety of foreign and domestic sources, including interviews with government officials and translations of foreign government documents. In most cases, multiple sources were consulted in order to verify information on foreign government programs.

8. This is an improvement over the original form of the tax credit in which the base year was a moving average of the previous three years of R&D spending. By fixing the years in which the R&D "intensity"(R&D-to-sales ratio) of the firm is determined, the increment on which the tax credit is calculated can grow over time if the firm becomes more R&D intensive.

9. These market failures are typically one of the following types: high intrinsic technical risk (quantum jump in the technology is being attempted, large financial risk [R&D is capital intensive and therefore expensive], long time to commercialization [market], and wide scope of potential markets [compared to existing corporate strategic foci]).

10. The Japanese government actually funds industrial R&D through eight different ministries/agencies, most of which have their own research institutes and quasi-government research corporations to distribute/manage/jointly conduct research.

11. The TRON Project did not involve government funds.

12. See, for example, Council on Competitiveness, 1991, p. 15.

13. The actual name is Key Technology Research Promotion Center.

14. Formally called "Framework Program for Research and Technology Development."

15. EUREKA is the European Cooperation on Advanced Technology Program.

16. The EC funds are allocated to the more generic elements of the EUREKA project.

17. It should be kept in mind that Japan's three leading agencies for R&D funding (MITI, MPT, and Ministry of Education) have stated their intention to double the Japanese government's funding of R&D by the year 2000.

REFERENCES

Aschauer, David A. "Does Public Capital Crowd Out Private Capital?" *Journal of Monetary Economics* (September 1989): 171–88.

Aschauer, David A. "Is Public Expenditure Productive?" *Journal of Monetary Economics* (1989): 177–200.

Aschauer, David A. "Rx for Productivity: Build Infrastructure." Chicago Fed Letter. September 1988.

Coe, David and Elhanan Helpman. "International R&D Spillovers." *European Economic Review*. 1995.

Coe, David, Elhanan Helpman and Alexander Hoffmaister. "North-South R&D Spillovers." *European Economic Review*. 1995.

Council on Competitiveness. "Competitiveness Index." Washington, D.C., 1995.

Council on Competitiveness. "Japanese Technology Policy: What's the Secret?" Washington, D.C., February 1991.

Erenburg, Sharon J. Linking Public Capital to Economic Performance. Annandale-on-Hudson, N.Y.: The Jerome Levy Economics Institute of Bard College, 1994.

Erenburg, Sharon J. "The Real Effects of Public Investment on Private Investment: A Rational Expectations Model." *Applied Economics* (1993): 831–37.

Justman, M. And M. Teubal. "Technological Infrastructure Policy (TIP): Creating Capabilities and Building Markets." *Research Policy* (1995): 259–81.

Leyden, Dennis and Albert Link. *Government's Role in Innovation*. Boston: Kluwer, 1992.

Mamuneas, Theofanis, and Ishaq Nadiri. Public R&D Policies and Cost Behavior of the US Manufacturing Industries. New York University, May 1993.

Nadiri, Ishaq. "Innovations and technological Spillovers" (NBER Working Paper No. 4423.) New York, N.Y. National Bureau of Economic Research, August 1993.

National Science Foundation. *Asia's New High-Tech Competitors*. Washington, D.C.: National Science Foundation, Science Resources Studies Division (NSF 95-309), 1995.

OECD (Organization for Economic Cooperation and Development). Science and Technology Policy: Review and Outlook. Paris, France. 1994

Tassey, Gregory. *The Impacts of Technology on Economic Growth: Implications for U.S. Infrastructure Policies*. Gaithersburg, Md.: National Institute of Standards and Technology, 1995.

———. *Technology Infrastructure and Competitive Position*. Norwell, Mass.: Kluver, 1992.

———. "The Functions of Technology Infrastructure and Competitive Economy." *Research Policy* (1991).

2

Regional Technology Infrastructures for Japanese Small- and Medium-Sized Enterprises: Policies and Programs for Technological Modernization

Philip P. Shapira

INTRODUCTION

Over the last few years, greater attention has been paid to the contribution of small and medium-sized firms to Japanese economic and technology development.[1] With more than eight hundred thousand manufacturing enterprises employing three hundred or fewer workers, Japan's small firm sector occupies more than three-quarters of the country's manufacturing workforce[2] (Table 2.1). The pressures on these small and medium manufacturing enterprises (SMEs) have grown as the Japanese economy has become more internationalized and as traditional larger customers change supply policies and sources. Japanese small manufacturers have been hard hit by the post-1991 collapse of the "bubble-economy" and the rising international value of the yen (making exporting more difficult, offshore sourcing more favorable, and importing more attractive even in Japan's difficult-to-enter markets). There has been a renewed debate in Japan about what was once viewed as mainly an American problem—the "hollowing-out" of the national manufacturing base, with the collapse of many small manufacturers and the weakening of local economies dependent on industrial production (Sumiya, 1994, p.1).

These developments have begun to modify the way the small firm sector is regarded by both policy makers and small firms themselves. There has long been a "dualistic view" about the position and potential of small- and medium-sized manufacturing enterprises (SMEs) in Japan. On the one hand, SMEs have been seen as technologically backward and distinct from Japan's successful large companies (with the latter traditionally receiving greater policy priority). Simultaneously, the tight vertical linkage between SMEs and their larger customers has often been identified as an important source of the technological

Table 2.1
Small- and Mid-Sized Enterprises, Japan 1991

	Enterprises 1991	Employment 1991
	Thousands	*Thousands*
Small- and Mid-Sized Enterprises of which:	852.3	10,396
under 20 employees	749.5	5,371
20–99 employees	90.9	3,517
Large Enterprises	4.6	3,691
Total	856.9	14,087
	Percent	*Percent*
Small- and Mid-Sized Enterprises of which:	99.5	76.6
under 20 employees	87.5	38.1
20–99 employees	10.6	25.0
Large Enterprises	0.5	23.4
Total	100.0	100.0

Note: Small- and mid-sized enterpries: Japan = less than 300 employees.
Sources: Small and Medium Enterprise Agency, *White Paper on Small and Medium Enterprises in Japan 1993*, Ministry of International Trade and Industry, Tokyo, 1993 (Appendix, Tables 1 and 2).

and economic strength of the Japanese industrial base. There remains much truth in each view. But both of these perspectives are now converging, if not giving way, to the idea that Japanese SMEs, bolstered through a variety of public and private relationships, can in the coming years assume an even more pivotal economic and technological role than in the past. Many Japanese SMEs now wish to reduce traditional vertical dependencies and build stronger horizontal and lateral ties with a wider variety of other enterprises and with research centers to secure business survival and greater technological autonomy. Japanese policy makers also want SMEs to assume a more prominent place in national and regional development (induced, in part, by the internationalization of the Japanese economy, which has largely taken the big firms out of the ambit of Tokyo's policy control). To support this direction, new regional technology initiatives focused at SMEs are being added to Japan's long-established programs for small firm technological modernization.

Figure 2.1
Small Manufacturing Establishments and Employment, Japan and United States

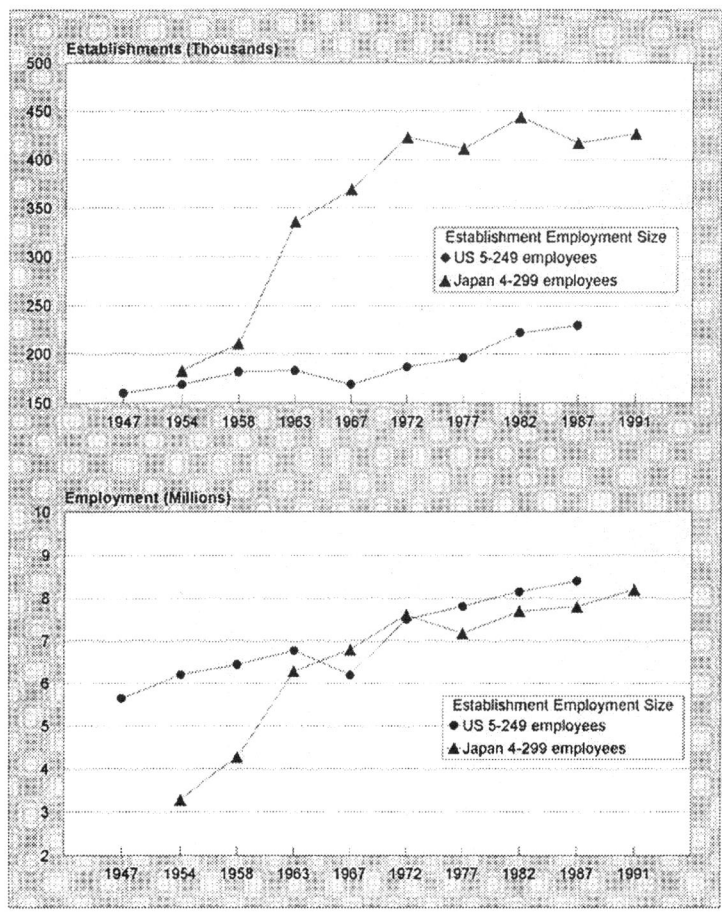

This chapter considers the changing position of SMEs in Japan and examines key elements of the public support system for technology promotion. The characteristics and operation of Japan's comprehensive system of local technology centers are reviewed, along with a discussion of new regional technology initiatives. The chapter concludes with an assessment of Japan's public policies and programs for SME technology promotion.

THE CHANGING POSITION OF SMEs IN JAPAN

Japan's small firm sector has grown significantly in terms of the number of establishments and jobs over recent decades (Figure 2.1). Today, Japan's 850,000 small manufacturers employ three quarters of Japan's manufacturing workers and produce more than half of the manufacturing value added. These small manufacturers are diverse.[3] About half are workshops with three or fewer workers, while the rest—still a huge segment of some 432,000 enterprises—each employ between 4 and 299 people. Often, the smallest workshops are engaged in traditional Japanese crafts, while other small firms are labor-intensive operations, producing simple components or carrying out routine tasks for larger companies. Some are labor-only subcontractors, with no plant or equipment of their own, who send personnel to work inside the plants of other companies. But many small firms are involved directly in modern manufacturing, using and developing advanced technologies. For example, the technological gap between small and large firms is smaller in Japan than in the United States, and small Japanese companies are more likely to use new technologies and techniques than their U.S. counterparts. Almost 50 percent more Japanese SMEs than U.S. SMEs use numerical or computer-numerical control machine tools, and they use more than four times as many advanced machining centers and robots (Shapira, 1992). Worker training—essential to the proper use of new technology—is also relatively strong in small Japanese firms (Office of Technology Assessment, 1990).

Japan's hundreds of thousands of small, flexible, and technologically proficient manufacturers are sources of high quality inputs and technological enhancements to larger companies. Contracting in Japan is typically organized in a pyramid fashion, with large manufacturers at the top supplied by smaller firms in multiple lower tiers. Long-term relationships between the tiers of smaller and bigger firms have given the smaller units the confidence to invest in new technology, workforce training, and ongoing product and process improvement (Dore, 1986; Trevor and Christie, 1988). But this "carrot" is usually accompanied by the "stick" of continuous improvement: most large companies require strenuous cost, quality, and delivery schedules, further driving smaller suppliers to modernize.

While many SMEs are closely tied to their larger customers, they also seek outside sources of support. Large firms may help their suppliers by sharing information, technology, and personnel—but not always. Sometimes, large companies may have little time to deal with the problems of smaller firms; or a small supplier may be more specialized than their larger customers, or may need special training or technological expertise. In such cases, small firms have to look beyond their larger customers for assistance. They may also seek outside help to deal with problems of adjustment. As large Japanese customers rationalize or internationalize production, many smaller firms are trying to develop new products to offset reductions in their traditional business lines. In other instances, large firms are themselves diversifying, compelling their smaller suppliers to shift into new technologies. Increasing competition from low-cost Asian suppliers, a highly valued yen, and the difficulty of attracting skilled young workers (who prefer larger firms) are also stimulating small firms to invest in new, labor-saving technologies and to upgrade working conditions. The bursting of Japan's "bubble-economy" of the late 1980s and the onset of a deep recession after 1991 has intensified the pressures on Japanese SMEs to rethink their technology and business linkages and strategies. Forced to slash costs and curtail production during the recession, major companies—especially in the hard-hit automotive and consumer electronics industries—have reduced or cut off orders from their smaller suppliers. This has occurred to such an extent as to call into question the future of Japan's tight parent company-supplier relationships. While these links are so embedded that any fundamental changes will take a long time, the recession has convinced many small enterprises not to rely so heavily on a single major buyer. Cost-cutting drives have led to a drop in SME capital investment, which in 1992 was down 9 percent from the level of 1988. SME business closures are up, while start-ups are down. At the same time, the recession appears to have opened new possibilities for SMEs with innovative technologies to exploit niche markets that larger companies are disinclined to enter. The current recession is thus giving further impetus to a trend that had been gathering momentum during the 1980s. An expanding segment of small Japanese manufacturers wants to develop, control, and sell their own products and technologies in domestic and international markets, without the constraints set by larger customers. One indication of this is the declining proportion of Japanese small manufacturers exclusively engaged in subcontracting, which dropped from two-thirds to just over one-half during the 1980s. The last decade saw an increase in R&D spending and personnel in many small firms and an improvement in design capabilities. Geographical clusters of small innovative firms have formed, in Tokyo and other large cities, and in less-urbanized locations such as the Nagano Prefecture in central Japan.[4] New horizontal and lateral relationships are being developed, between and among small firms and in joint ventures with larger firms. A major aim of recent Japanese local and regional public technology policy is to reinforce and stimulate these emerging SME business development strategies.

MODERNIZATION POLICY IN JAPAN

There has been an important evolution in Japan's policies toward small firms (those with three hundred or fewer employees). In the years after World War II, small firms were often viewed as a backward sector in Japan. As policy makers favored building up large firms, especially in heavy and mass production industries, efforts were made to combine some small firms into bigger ones, and there has been an important evolution in Japan's policies toward small firms (those with three hundred or fewer employees). In the years after World War II, small firms were often viewed as a backward sector in Japan. As policy makers favored building up large firms, especially in heavy and mass production industries, efforts were made to combine some small firms into bigger ones, and to shelter the rest from more efficient competition. More recently, while some protective measures still exist, the thrust of Japanese small firm policy has been to stimulate upgrading and modernization. Fostering innovative, knowledge-intensive small firms is considered vital to Japan's shift toward high technology and more flexible production methods. Small enterprise promotion and technology improvement are also seen as important regional economic development tools.

Modernization in Japan means not only strengthening technology, facilities, management, operations, and human resources in small firms themselves, but also improving entire sections of small enterprises, including inter-company and inter-industry relationships. Japan's national small enterprise laws and policies establish general mechanisms to provide finance, tax incentives, guidance, and assistance to individual companies and groups of small firms. There are also special measures to help small companies convert to new business lines and develop new products and technologies. The national government assists local programs that provide small firms with technology development support and guidance. Funds are made available for local technology centers, the clustering of companies for joint product development, marketing, training, improvements in design abilities, information dissemination, and research and entrepreneurship.

The principal central agency responsible for small firms in Japan is MITI's Small and Medium Enterprise Agency (SMEA), whose functions include overseeing small industry guidance and technology development, subcontracting, enterprise promotion, and planning and research. Separate national councils for small and medium enterprise policymaking, stabilization, and modernization provide advice and review. Other agencies and bureaus contribute, including MITI's Machinery and Information Industries Bureau and the Agency of International Science and Technology (attached to MITI). An associated public organization, the Japan Small Business Corporation, provides guidance and financing for structural improvement and upgrading projects in small firms, training for enterprise personnel and local program staff, information and computing support, and business finance. The national government has also established three major financial institutions, targeted at small- and medium-sized firms, to complement private financing and promote specific modernization policy goals: the People's Finance Corporation, which extends funds to very small-scale firms; the Small Business Finance Corporation, which supplies longer-term funds to small and mid-sized firms, and the Shoko Chukin Bank,

Regional Technology Infrastructures

which finances small firm cooperatives and small industry organizations[5] (Figure 2.2).

A recent national policy encourages small firms with different specialties to work together to develop and commercialize new products.[6] This is carried out by establishing local technology plazas or meeting places, supporting mediators, sharing information, offering subsidies to business associations and fusion groups, and providing support for shared production and marketing. By 1994, about two thousand five hundred SME fusion groups had been registered in Japan. Most of these were still at the stage of initial association and research, but some had moved to commercializing jointly developed new products (Shapira, 1994, p. 10–12).

At the local level, there are prefectural and city offices of industry promotion, which includes local small firm development and guidance. Equipment modernization loans and leasing systems for general, high technology, and information processing equipment are available for small firms, funded jointly by national and prefectural governments. Prefectures and cities also make additional funds and incentives available to local small firms for plant and equipment investments, through tax relief, interest subsidies on private bank loans, and other allowances. Area business and industrial associations play an important role in small industry modernization in Japan, a role supported by the public sector. Local chambers of commerce, industry federations, subcontractor promotion associations, and industry-specific structural improvement associations all receive public financial support. An example of the integration of several of these programs is found in Ehime Prefecture, on the southern island of Shikoku, where a Towel Industry Resource Center is working to revitalize local small towel firms, by promoting new computer-aided design (CAD), manufacturing methods, design consulting, training, and joint marketing. This center is a cooperative effort involving the local industry association, the prefectural and city governments, and MITI. Next door, programmers, CAD operators, and information specialists are trained in a new Computer College established with a foundation grant from the Ministry of Labor's Employment Promotion Corporation.[7]

The variety of programs in Japan, together with the fact that small firm industrial modernization initiatives overlap with programs for general business development, regional development, and technology promotion, makes it difficult to precisely calculate Japanese spending for small-firm technology assistance. Only a small portion of small business assistance is counted in the regular central government tax budget; most central resources are provided from trust funds and other capital accounts. The U.S. Office of Technology Assessment reckons that more than US$30 billion (ECU 24 billion)—or more than 5 percent of the national regular and capital budgets in Japan—goes to support small firms each year, including non-manufacturers and loans, but ex-

Figure 2.2
Japan's Public Support System for SME Technology Promotion

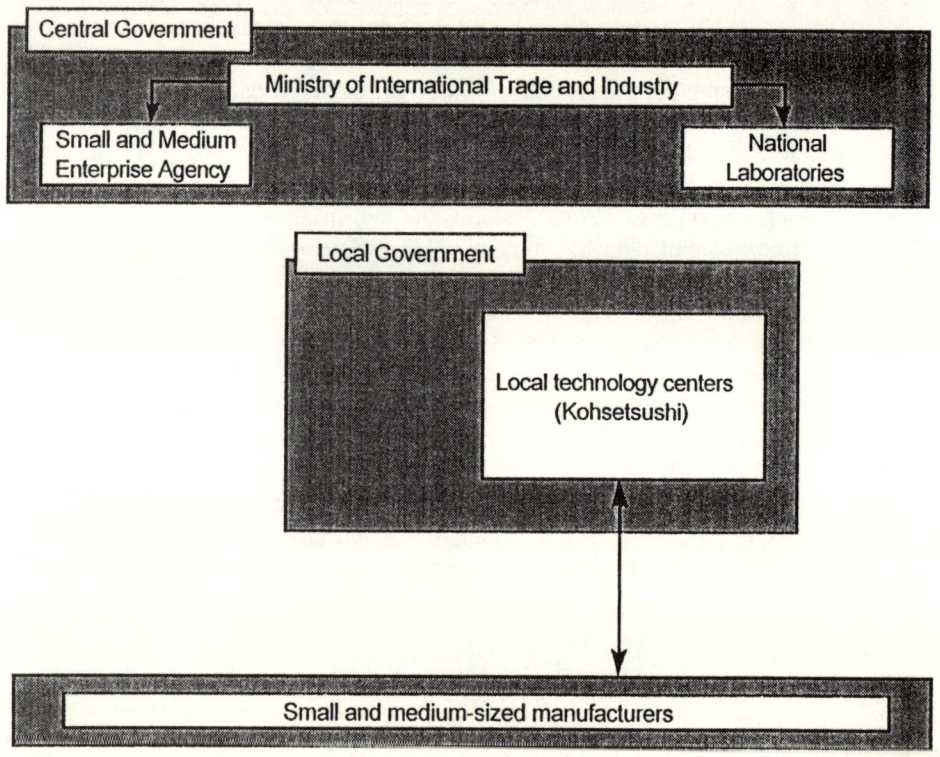

Figure 2.3
Kohsetsushi Centers and Regional Technology Projects in Japan

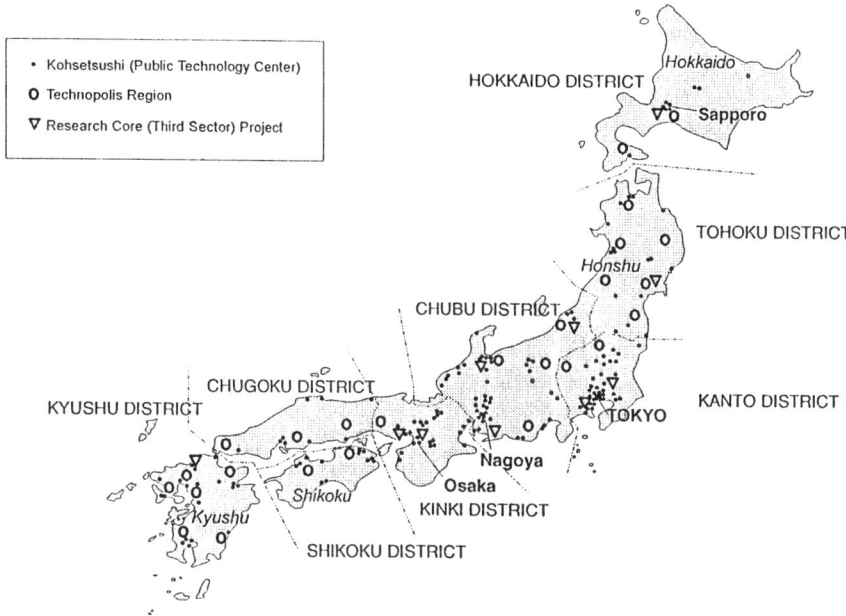

cluding prefectural and city contributions and spending on related regional development and technology programs (Office of Technology Assessment, 1990).

KOHSETSUSHI CENTERS

A cornerstone of Japan's system for modernizing small firms is its nationwide network of local and prefectural technology and testing centers. Known as Kohsetsushi centers (an acronym for *koh*, public; *setsuritsu*, establishment; and *shikenjo*, testing laboratory), these are publicly sponsored institutions, with very large engineering staffs, that serve as free or almost free resources for manufacturers with three hundred workers or less.[8] Japan began to establish these industrial research, experiment, and testing institutes at the turn of the twentieth century, based in part on the U.S. model of agricultural experiment stations and extension large services. National and university institutes were the first to be founded, followed in the 1920s and 1930s by a number of local government centers to strengthen local industries. After World War II, additional local Kohsetsushi centers were constituted. In recent years, a few new centers have been added, while many older Kohsetsushi centers have expanded or built new facilities.

Today, there is at least one center in each of Japan's forty seven prefectures, with 22 centers in the Tokyo metropolitan region[9] (Figure 2.3). In total, there

Table 2.2
Japan's Public Technology Centers Measures of Performance

	Kohsetsushi Centers[1] 1994 Annual Rate
Centers	178
Total Funding	¥98.16 billion (US$ 1.0 billion; ECU 812.6 million)
National funding	10-20%
Staffing	5,200 (engineers) 6,800 (with support personnel).
Staff Research Time	30-50%
Technological Assists[2] (annual cases)	All contacts: 406,000 Field visits: 11,300
Training seminars and workshops held	7,800
Inspections/Exams	684,000
Open labs/Demonstrations	63,500
Technology Diffusion Groups/Networks[3]	2,500

[1]Date provided by Ministry of International Trade and Industry, Tokyo. Service data is for 1991. Currency converted at market exchange rates of US$ 1.00 = ¥ 98.21 and ECU 1.00 = ¥ 120.8. These exchange rates do not reflect differences in purchasing power.
[2]Definition of technological assists in Japan is broad in Japan (includes any type of contact).
[3]The estimate of technology diffusion and networking groups includes groups sponsored by other organizations.

are now about 180 centers, employing six thousand eight hundred people, including five thousand two hundred engineers and technical personnel (Table 2.2). The centers usually develop expertise in technologies used by local industries, with each center maintaining several technological specialties. The greatest weight is in traditional craft industries, chemicals, metalworking, food processing/biotechnology, electrical and electronics engineering, textiles and clothing, ceramics, and distilling (Figure 2.4). About half of the centers employ under twenty nine staff members each. But there is a group of about forty bigger centers, with fifty or more staff members, mostly found in urban areas. The largest centers—in Tokyo, Osaka, Nagoya, and other big cities—each have more than one hundred staff members.

The Kohsetsushi centers are administered by prefectural and municipal governments, who also provide most of the funding. The overall Kohsetsushi budget was ¥98.16 billion ($US 999.5 million or ECU 812.6 million) in FY 1994. Typically, the central government provides about 10 to 20 percent of the financing for each center, with funds coming from MITI, the Japan Small Business Corporation, and the Japan Bicycle Development Association (which

Figure 2.4
Kohsetsushi Centers Technological Specialties and Size of Staff

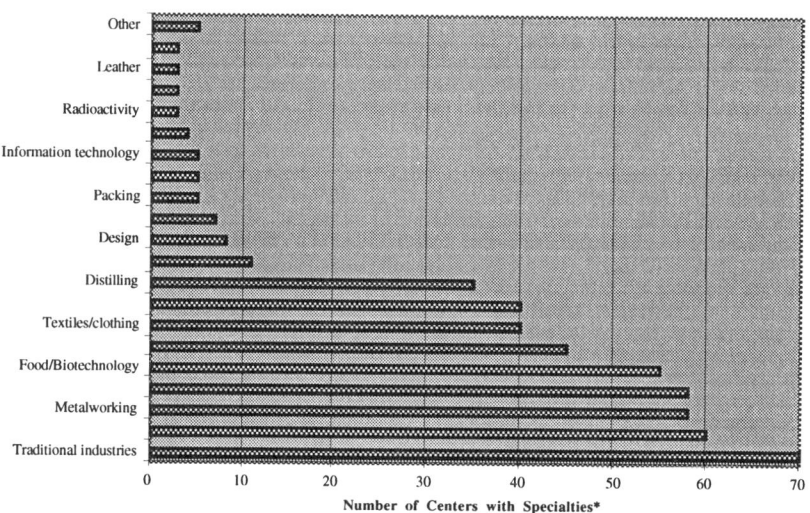

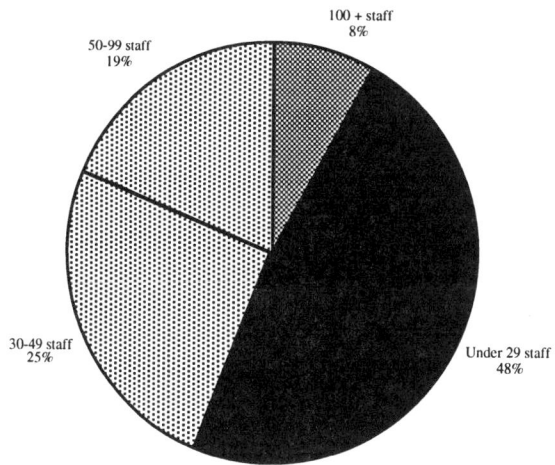

uses betting profits from bicycle racing to improve machinery and metalworking industries). Fee income from services to private firms is small.

Kohsetsushi staff spend up to half their time on research, mainly on applied projects focused toward local industries. Projects may be organized by the center itself, or sponsored with local companies or universities. Small manufacturers often send one or two of their staff to work on Kohsetsushi research projects, providing opportunities for company personnel to gain research experience, develop new technical skills, and transfer technology back to their firms. The centers run a variety of seminars and study meetings to disseminate information on research and new technologies to local firms, as well as publish newsletters and research reports and maintain technical libraries.[10]

Conducting tests and examinations is another major Kohsetsushi activity. For nominal fees, Kohsetsushi laboratories will analyze materials and products, verify standards compliance, calibrate measuring instruments, and make sophisticated testing equipment available. These services are much used by Japan's small firms, with more than 680,000 tests and examinations conducted yearly. These services help small manufacturers to enhance quality, precision, and product development, in addition to resolving problems in materials and components.

To help small companies overcome technical difficulties and implement new technology, Kohsetsushi centers provide advice and guidance services. For simple requests, enterprise managers call in by telephone or visit the center; more complex problems are dealt with by field visits from center staff to companies. Over 400,000 contacts are handled annually, including 318,000 cases where technological consulting is provided (usually at the center), and over 21,000 instances where staff teams or advisers visited firms. About half of these visits are made through a program of technology advisers whereby Kohsetsushi centers match company needs and requests with registered private manufacturing consultants. The advisers are initially reimbursed from local and central funds, allowing them to provide a first round of services at no cost to firms.

Training in new technologies for employees of local small manufacturers is provided through Kohsetsushi group and customized programs. Employees go to the centers for classroom instruction and hands-on experience with advanced tools, computers, and software systems. Many Kohsetsushi centers offer open laboratories, making their specialized equipment available for research, prototyping, and training. Kohsetsushi centers also sponsor technology diffusion and network groups to encourage small firms to exchange information, share technology, and develop new products and markets. Each center may sponsor several such groups, with each involving up to thirty local firms.

NEW REGIONAL TECHNOLOGY PROJECTS AND PARTNERSHIPS

Complementing the long-established Kohsetsushi centers is a newer, and still growing, infrastructure of regional technology projects, industry resource centers, and technocenters. Concerned with lagging regional development outside of Japan's booming Tokyo-Nagoya-Osaka central core, MITI and other central government ministries have established a series of projects to promote the technological upgrading of existing industries and the development of new technologically based enterprises in other regions.

One of the most well-known of these initiatives is the Technopolis program, which, in the 1980s, designated twenty six areas to serve as nodes for high technology growth.[11] At many of these Technopolis sites, new technocenters have been built, which are helping to introduce advanced technologies to local small firms. Regional technology development is also the aim of the Research Core program, which equips special facilities for promoting small firm technology transfer, business incubation, and training (Figure 2.5). Sponsored by MITI, although funded mainly by local government and the private sector, ten Research Core locations have been chosen to date.[12] Other regional technology projects championed by MITI and other ministries include the Key Facilities Concept (promoting facilities for information services and research in peripheral areas) and the New Media Community (developing new information systems in part for local firm networking).[13]

These regional technology initiatives are complemented by an expanding group of new local industry resource centers, city technocenters, and training institutes. For example, on the northern island of Hokkaido, the Muroran Technocenter is providing area small manufacturers with assistance in training, consulting, marketing, networking, information distribution, research and development, and the open use of advanced machines to help diversify the heavy industrial base of the local economy.[14]

Many of the new regional technology projects and industry centers are structured as "third sector" organizations. This usually involves a governing foundation comprised of public and private representatives, which allows more flexibility in activities and staffing, and enables resources to be leveraged from the private sector. Funding comes from private member companies, banks, utilities, and local and prefectural governments. The national government contributes equity capital and loans for third sector organizations through an Industrial Infrastructure Improvement Foundation, which is endowed with proceedings from the privatization of the telephone company NTT.

Japan's third sector approach borrows from the public-private partnership models developed in the United States. Policymakers have promoted the concept of the third sector in Japan to allow greater flexibility and autonomy than allowed in purely governmental operations. However, the desired effects have yet to be felt. Most third sector organizations use seconded government personnel who employ management systems and methods little different than

Figure 2.5
Research Core Concept

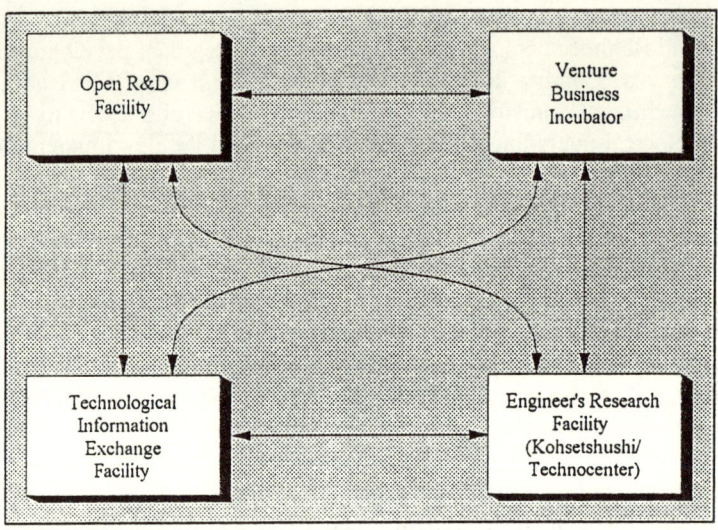

those found in the public sector. Moreover, during Japan's booming "bubble-economy" era, numerous third sector organizations built (or were provided with) lavish showcase facilities.[15] In the current recession, these buildings are costly to operate and hard to fill with revenue-generating activities. Further private contributions are not easy to obtain, which has forced national and local governments to supply additional soft subsidies. Good young researchers are also said to have some reluctance in joining third sector technology organizations because of the fear of instability and the lack of center reputation. At the same time, it must be added that most third sector organizations are at an early stage of development. Their primary sponsors show few signs of retreat and are likely to persist in providing support to help these organizations build capacity and become more effective.

ASSESSMENT OF JAPANESE APPROACHES

The combination of Kohsetsushi centers, regional and local technology projects, financial incentives, and other national and local policies provides an extensive support system for Japan's small firms. Most small firms have easy access to these assistance resources. The emphasis on examination, testing, and analysis in the Kohsetsushi centers has been valuable in helping small firms meet high standards of quality, performance, and precision. The ready access to advice and guidance services and advanced equipment, training, and information services assists firms in upgrading their operations, products, workforce, and strategies. And the centers help companies establish collaborative local research and act as a bridge to national research laboratories and universities (Table 2.3 provides a summary characterization of the Kohsetsushi centers and regional technology projects).

The Kohsetsushi centers present an intriguing illustration of nationally standardized services, but with little direct national funding. Across the country, different Kohsetsushi centers offer a remarkably standardized range of services and programs. They pursue the same policies and offer similar kinds of assistance to firms. Typically, the centers allocate about one-half of staff time to research, with the balance divided between (a) examinations and analysis, and (b) technology advice and guidance. Additionally, most centers sponsor technology diffusion groups or "plazas," in conformance with MITI's emphasis on this strategy.

This harmonization of services is, of course, attributable to MITI, which has a great influence on the Kohsetsushi system even though its direct funding share is small. There are several mechanisms through which MITI achieves this influence: first, by exercising "administrative guidance" *(gyosei shido)*—the extra-legal means of obtaining adherence to policies and practices deemed desirable by the government, which is longstanding and prevalent in (although not unique to) Japan; second, through the linkages between the Kohsetsushi centers and MITI's national research laboratories, which guide Kohsetsushi research approaches; third, through personal connections and personnel rotation, since many government officials at the local level (both elected and appointed) have links with MITI or are former MITI employees; and fourth, through carefully leveraging small amounts of MITI project funding for new research projects, training programs, and technology assistance and diffusion activities and funding from MITI-affiliated bodies such as the Japan Bicycle Development Association. Japan's centralized and bureaucratically led policy apparatus readily adopts a long-run view, although the infighting between different ministries and the stifling of local program experimentation, innovation, and flexibility by centralized control are considerable weaknesses.

Japanese firms often turn to the Kohsetsushi centers for aid in improving existing technologies and products. Companies report that local technology centers are helpful here and that the free or nominal cost of assistance encourages them to use the public services. In most cases, center staff are able to assist in these everyday problems, which usually do not require the latest technological expertise. Noting this, some Japanese observers criticize the weak quality of Kohsetsushi research and technology. Visits made to centers indicate a measure of truth in this view. In some instances, laboratories are visibly decayed, underused, and full of old equipment; many projects are run by single researchers who also have many other responsibilities; the average age of researchers is high; and the titles of some published papers are mundane (e.g., examining PC use in manufacturing enterprises). In other cases, Kohsetsushi

Table 2.3
Kohsetsushi Centers and Regional Technology Projects in Japan

	Kohsetsushi Centers	Regional Technology Centers
Sponsors	Prefectures and large cities	Prefectures and cities, with the private sector, through public-private partnerships
Primary users	Small and medium enterprises, 300 or fewer employees	Small and medium enterprises, 300 or fewer employees; large enterprises
Services	Research; technical advice and guidance; examination and analysis; information dissemination; training; open laboratories and equipment use; registered technical advisers; diffusion of technology groups	Research; technical advice and guidance; information dissemination; training; venture business support; research facilities for existing firms; open laboratories and equipment use; technology exchange and diffusion groups
Coverage and standardization of services	Nationwide coverage; standardized services; almost no experimentation in services	Numerous selected sites, mainly outside Tokyo and Osaka; similar services; small amount of experimentation
Staffing	Life-time staffing, little flexibility; almost all staff have engineering or technical qualifications	Dispatched staff from public and private sector
Research links	Centers maintain own research programs; links with national labs; some university links	Most have own research programs; links with local universities, labs
Service delivery methods	Generally informal; large number of small interactions; some cooperative research projects with SMEs no or nominal fees for service	Make high quality facilities and equipment available; provision of advice and guidance; some additional subsidies
Program evaluation	No formal evaluation systems	No formal evaluation systems
National role	Some national funds; centralized guidance over local activities; determination of overall program direction; identification of key technology priorities	Some national funds; centralized guidance over local activities; determination of overall program direction
Program status	Mature but may soon be restructured; funding stable	Developing; massive expansion of facilities in "bubble" era; some funding shortfalls
Key issues	Lack of flexibility; staff expertise; promotion of innovation; level of technology; relationship to new regional third-sector initiatives	Difficulty of attracting good technical personnel; "soft" systems; flexibility; barriers to new technology start-ups; quality of research

research laboratories are well organized and equipped, and focused on ambitious research objectives. The general impression, however, is that most Kohsetsushi research is "catch-up" rather than pioneering.

Paradoxically, this "weakness" in research could be considered a strength of the Kohsetsushi system. Kohsetsushi researchers spend a lot of their time catching-up on work done in national laboratories and other research centers. Some Kohsetsushi staff report that, with their other responsibilities for guidance and assistance, it is a continual challenge to improve their knowledge about developments in their fields. In this "catch-up" mode, Kohsetsushi staff are not at the leading edge of their fields. Rather, they are in an intermediate or broker position where they are not too far ahead of their small firm clients. In the past, this meant they could readily transfer knowledge to smaller firms that was useful and close to applications. However, there is an increasing concern that a growing number of Japanese SMEs have technological capabilities well ahead of those found in most Kohsetsushi centers, leading to calls to upgrade the levels of center research and technology.

In seeking to assist and foster these more innovative small firms, many local governments want to increase the role of local centers in innovative technologies by funding more future-oriented research. New third sector institutions focused on advanced technologies have also been built. However, while the physical infrastructure for these new technology centers is impressive, there is a great weakness in the "soft" infrastructure of personnel and operating procedures to support innovation. Japan's rigid labor market makes it hard to attract talented young technologists to centers that have yet to establish a reputation, and professionals with industrial experience are unable to leave their current employers in mid-career. Additionally, despite their intent, most third-sector centers are managed by dispatched public sector personnel, effectively transferring in the very practices of inflexibility and risk aversion these centers were designed to overcome.

Indeed, staffing has emerged as a major issue for most of Japan's local technology centers. The Japanese have tended to use career staff, generally with university engineering or science undergraduate degrees, to provide core services in the Kohsetsushi centers. Usually, the staff work their whole career with the sponsoring prefecture, often in a single center. This ensures stability and helps build long-term relationships with local firms, but staff skills can become outdated, and the low turnover limits opportunities to recruit young staff in new areas of technology. This is a problem now in sharp focus as the Kohsetsushi centers seek to advance the technology frontier of their research. Kohsetsushi centers are trying to address this question by hosting visiting researchers and increasing education and training for existing staff. But the Kohsetsushi centers (along with the new third-sector technology centers) continue to find it difficult to attract the best young researchers and technical staff in areas of new technology.

In the future, the policies of both central and local government to actively promote technology upgrading in small and medium manufacturers will lead to additional emphasis, resources, and demands for technology services providers at the local level in Japan. In some instances, Kohsetsushi centers are facing competition from the latest generation of regional technology programs. More frequently, Kohsetsushi centers are working with new third-sector initiatives in regional technology partnerships. However, the growing variety of regional technology schemes in Japan still presents issues of coordination of resources and, most critically, ones of relevance and effectiveness. On the latter point, the concern is whether in a fast-changing technological and business environment, Japan's public and public–private modernization systems can be sufficiently flexible, targeted, innovative, and customer driven to meet the changing needs of the small manufacturing base.

In an acknowledgment of the challenges facing the Kohsetsushi system, the Small and Medium Enterprise Agency has established a Technology Policy Committee to provide advice about new legislation for SME technology promotion and the restructuring and future role of the Kohsetsushi centers. This effort will contribute to a new national "vision for the 21st century for technology policies for small and medium sized businesses," set to be announced by 1995.[16] It is likely that this vision will seek to shift the Kohsetsushi centers away from testing and technology guidance, to focus more on advanced research in fundamental and applied industrial fields. This research will be related to regional needs and aims to nurture the technological strengths of SMEs. Preparing the ground for this new direction, a regional study committee has already recommended the improved integration of Kohsetsushi centers in regional industrial policies, upgraded technological capabilities within the centers, wider research collaboration with other institutions and companies, improved researcher training, and a more active role in working with local companies.[17] One official has suggested that the Kohsetsushi center "in 10 years will probably have a completely different image than it has today."[18]

Japan's policy commitment to technological modernization is strong, and there is more robust political and financial support for a comprehensive system. The range of information, technical, and assistance services provided to small firms is marked by two key characteristics: (1) remarkable standardization, with little variation from place to place; and (2) a strong engineering and hard technology focus. There is little variation or experimentation in program services between different localities, and formal evaluation systems are practically nonexistent. This system has worked effectively in the past, but as the technological and business environment changes, Japan now seeks to find ways to evolve its system into one that is more innovative, flexible, and decentralized. New programmatic elements are being added to address the changing needs of SMEs. It remains to be seen, however, whether Japan's Kohsetsushi and regional technology centers can implement new research approaches, organizational styles, and personnel systems to make them fully effective in new technology development and avoid unnecessary research duplication. This is an exacting challenge, which will require rather fundamental and difficult alterations in the way Japan's public technology programs are structured and operated in the future.

NOTES

This chapter draws on research supported by the Georgia Tech Foundation, the Japan Institute of Labour, the Fraunhoffer Gesellschaft Institut für Systemtechnik und Innovationsforschung, and the Center for International Business Education and Research at Georgia Institute of Technology. The views expressed are those of the author and do not necessarily reflect those of these research sponsors. Nominal currency exchange rates of $U.S. 1.00 = ¥ 98.21 and ECU 1.00 = ¥ 120.8 are used here.

1. See, for example, D. Friedman, *The Misunderstood Miracle: Industrial Development and Political Change in Japan*, (Ithaca, N.Y.:, Cornell University Press, 1988).

2. Japan's Small and Medium Enterprise Law defines a "small and medium enterprise" in manufacturing as one with three hundred or fewer employees or with capital of ¥100 million (US$ 1.02 million or ECU 0.83 million) or less. A "small-scale enterprise" in manufacturing is defined as one with twenty employees or less.

3. See, for example, N. Chalmers, *Industrial Relations in Japan: The Peripheral Workforce*, (London and New York: Routledge, 1989).

4. See, for example, D. Friedman, *The Misunderstood Miracle: Industrial Development and Political Change in Japan* (Ithaca N.Y.: Cornell University Press, 1988).

5. For discussion of Japanese policies for small firms, see: Small and Medium Enterprise Agency, *Small Business in Japan 1994*, White Paper on Small and Medium Enterprises in Japan, Ministry of International Trade and Industry, Tokyo (and earlier issues of this annual publication).

6. See Extra-Ordinary Law Concerning the Promotion of the Development of New Business Areas Through Fusion of Knowledge of Small and Medium Enterprises in Different Industries ("Fusion" Law), 1988.

7. Interviews conducted with Shikoku Towel Industry Association and Ehime Prefectural Local Industry Promotion Center, Imabari, Ehime Prefecture, February 23, 1993.

8. The discussion of Japan's Kohsetsushi centers draws on field research in Japan in 1989, 1990, 1993, and 1994. The first part of this research is reported in greater detail in P. Shapira, "Modernizing small manufacturers in Japan," 1992, *op. cit.*; and P. Shapira, "Helping manufacturers do better: Japan looks after the little guys," *IEEE Spectrum*, 30, 9, September 1993. For subsequent field research by this paper's author on Kohsetsushi centers (and other Japanese regional technology policies), see: P. Shapira, *1994 Modernization Forum Study Mission to Japan*, Dearborn, Mich.: The Modernization Forum, 1995; and P. Shapira, (Tokyo: *Restructuring and Employment Development in Japan's Regions*, Japan Institute of Labor, 1995).

9. For a full and updated listing of Kohsetsushi centers, see *Public R&D Institutes of Local Governments in Japan* (Tokyo: Japan External Trade Organization [with the

Small and Medium Enterprise Agency and the Agency of Industrial Science and Technology], March 1991.)

10. For a detailed description of a Kohsetsushi center—the Tokyo Metropolitan Industrial Technology Center—see: P. Shapira, *Technology Development and Application Centers: Case Study Examples from the United States and Japan*, (Atlanta: School of Public Policy, Georgia Institute of Technology, 1994).

11. See I. Masser, "Technology and regional development policy: A review of Japan's Technopolis programme," *Regional Studies*, 24.1, 1990, pp. 41–53.

12. An example of a Research Core Project—the Kanagawa Science Park—is detailed in Shapira, 1994.

13. See D.W. Edgington, "New strategies for technology development and information systems in Japanese cities and regions," in P. Shapira., I. Masser, and D. Edgington, ed., *Planning for Cities and Regions in Japan*, (Liverpool, U.K.: Liverpool University Press, 1995).

14. Interview and field visit, Muroran Technocenter, Muroran, Hokkaido, January 26, 1993.

15. It is worth noting that Japan's political and budgetary systems often result, first, in the construction of a capital facility, with funding for operations gradually being allocated afterward. In this sense, the third sector centers are not unusual, although it confirms the point that they are still being run under traditional public sector procedures.

16. "MITI to formulate 'Technology Vision' for small, midsized enterprises," *Nikkan Kogyo Shimbun*, June 29, 1993.

17. Kinki Bureau of Trade and Industry, *Reinforcement of Small and Medium Companies*, (Osaka: Report of the Committee on Kohsetsushi and Technology Reinforcement of Small and Medium Companies, 1994).

18. "Chugoku GIRI chief calls for restructuring of Kohsetsushi," *Tokyo JITA News*, 93FE08111, June 1993, pp. 2–3.

REFERENCES

Dore, R. *Flexible Rigidities*. London, England: The Athlone Press, 1986.

Shapira, P. "Collaborative business exchange and technology fusion: The Japanese Approach". *Firm Connections*. September/October 1994, pp. 10–12.

———. "Modernizing Small Manufacturers in Japan: The Role of Local Public Technology Centers." *Journal of Technology Transfer*. Winter 1992.

Sumiya, F. "Small Manufacturers Face Survival Fight." *Nikkei Weekly*. June 13, 1994, pp. 1.

Trevor, M., and I. Christie. *Manufacturers and Suppliers in Britain and Japan*. London, England: Policy Studies Institute, 1988.

U.S. Congress, Office of Technology Assessment. *Making Things Better: Competing in Manufacturing*. Washington, DC: US Government Printing Office, 1990.

3

Living on the Edge: Basic Research and Knowledge Creation in Japanese Electronics Companies

David T. Methé

One of the most important aspects of a nation's technoinfrastructure concerns the balance between the process of creating new knowledge and the process of disseminating this knowledge. In modern societies these tasks are assigned to distinct institutions such as universities, government research laboratories, and business laboratories. Japan, partly by historical accident and partly by design, has tended to emphasize the dissemination aspect. This situation was facilitated by the fact that new knowledge could be imported and modified to meet the needs of Japanese society. This situation is now changing. Knowledge created elsewhere is no longer as readily available or as easy to import. Consequently, Japanese institutions are now confronted with the necessity of creating new technological and scientific knowledge, if they wish to participate in setting technical standards for products in the future. In this chapter, we will examine how one set of institutions, Japanese electronic companies, is coping with this new technological imperative. These attempts must be integrated into an already ongoing R&D system with established routines for carrying out product and process innovation. We will first examine this system and determine the most important features of it and the environmental forces acting on it for change. Then we will explore how ten Japanese electronic companies are implementing changes in their R&D systems, and what some of the implications of these changes are.

DOMINANT LOGIC OF COMPANIES IN JAPAN

The concept of a dominant logic requires that an organization be confronted with an environment that presents the organization with a consistent set of circumstances over time. In reaction to this, the organization makes a consistent set of strategic decisions, which in turn self-perpetuate themselves in the value orientation of managers in the organization (Prahallad and Bettis, 1986). Such a case emerged in Japan over the roughly forty years from the end of World War II until now. The Japanese economy was characterized by rapid economic growth. Various sectors of the Japanese economy, beginning with steel and shipbuilding and moving to automobiles and electronics, often grew at a double digit pace. Managers confronted with this situation developed "a winner's competitive cycle" mentality (Abegglen and Stalk, 1985). Market share became the overriding strategic goal, which meant that companies had to match or exceed the growth rate of the industry they were in if they wished to maintain their market share position. Maintaining market share was crucial, for as the growth rate slowed in one segment of the industry, the leading firms would reap the profits. Consequently, a dynamic of overinvestment to meet anticipated future market growth was created. Those companies that invested in plant and equipment to produce new products would gain market share at the expense of those that did not.

This dominant strategic logic impacted all aspects of the value chain. Our concern in this chapter are the activities related to research and development. Organizations also have a dominant technical logic (Lane, Beddows, and Lawrence, 1981). Again, this logic relates to the overall concept of how an organization puts together the necessary resources to carry out the research and development function. Fundamentally, organizations carrying out research and development must find a balance between a set of activities that exploits existing knowledge bases and a set of activities that explores and creates new knowledge bases. Exploration emphasizes the act of discovery, that is, the finding of knowledge for the first time, whereas exploitation emphasizes the act of invention, that is, utilization of existing knowledge to make something new. Although it is somewhat simple to characterize the research and development process in this way, the act of discovery or exploration is often termed "basic" research, while the act of invention is termed "applied" research and development.

The institutional arrangements for research facilities in Japanese companies can be seen as an outgrowth of the dominant strategic logic and its' influence on the dominant technical logic approach to research and development that already exists in these companies. The knowledge exploration approach most often discussed in the United States, of basic research providing the fundamental knowledge that applied research then turns into a prototype, which can be further refined into a successful product through developmental research, is turned on its head in Japanese companies. The Japanese "pull on the string" of existing knowledge and exploit what is readily available (Methé, 1995). It should be noted that this approach is not something inherent in Japanese culture or value systems, but it did grow out of the set of

Figure 3.1
Comparison of Knowledge Exploration and Knowledge Exploitation Approaches toward Research and Development in Companies

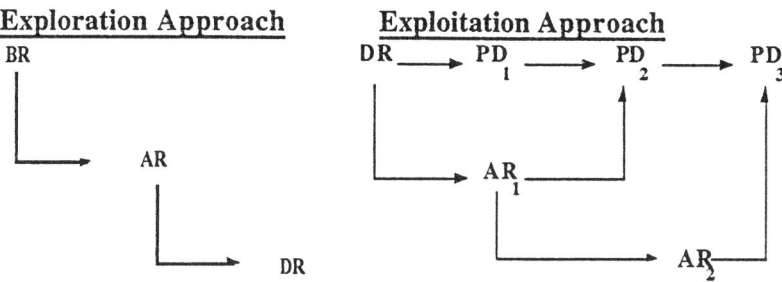

Note: BR = Basic Research; AR_i = Applied Research; Pd_i = Product Developed at i; DR = Developmental Research

environmental circumstances in which Japanese firms found themselves over a period of some forty years. As a late industrializing nation, knowledge was available, and the cost of *making* it was too high relative to the cost of *buying* it. It should be further noted that this knowledge exploitation approach is utilized by many American firms, such as Intel, which spend little or nothing on basic research (Figure 3.1).

As noted before, this approach of "pulling on the string" was well suited to the market share-driven strategy of Japanese companies, because it allowed a large variety of products to be developed from a well-mapped body of technical knowledge in a very efficient manner. The dominant technical logic of this research and development system places a larger emphasis on development and a smaller emphasis on research, because the knowledge needed was almost always available or acquirable. The acquisition of knowledge in this approach is problem driven. In order to meet a particular market need by a product that is often a combination of technologies, knowledge about those technologies must be accessed. Given the dominant strategic logic of rapid market share growth, that knowledge had to be accessed efficiently. Over time, Japanese companies were able to develop organizational routines, which facilitated this efficient accessing and exploitation of knowledge. This, in turn, led to the ability of Japanese companies to turn out large varieties of products (Table 3.1).

What can be observed is the number of new product introductions each year for a selected group of consumer electronic and durable white goods. The timing of the product introductions coincided with the semiannual payment of bonuses. The variety of new product introductions increased until 1989, and although still large in the remaining years, has begun to fall. We will return to

Table 3.1
Number of New Product Introductions,
Selected Products, 1987–1991

	1987	1988	1989	1990	1991
Color TV	277	286	296	223	189
VCR	154	202	184	150	99
Personal Computer	135	214	166	72	60
Air Conditioner	337	452	536	520	424
Refrigerator	189	235	228	196	151
Washing Machine	115	145	145	82	70

Source: Adapted from Nikkei Electronics, 1992.

this again. First, it is important to understand the dynamic underlying this product proliferation.

By examining the product introduction strategy of one company with one product, we can see how this dominant technology works. The product family is made up of several categories of related products (Figure 3.2). In this case, they are a "High End," "Main," "Compact," "Laptop," and "Notebook" series of products. Each of the circles represents a new model introduction, and does not include variations off the model, such as types of monitors, or number and combination of hard and floppy disk drives. What relates these various series of products into a family is that they all share some common technology, such as microprocessor or hard disk drive or floppy disk drive across each of the series.

The High End series generally was the first to exhibit the latest technology, in terms of microprocessor and hard disk drive capacity. These products were usually expensive, and would be followed by a more moderately priced product embodying the same technology, but at a later time of entry. The "Main" series are desktop personal computers that offered 5.25 inch floppy disk drives (FDD). When 3.5 inch FDD became available, these were incorporated in a new series, the *Compact*, which was eventually discontinued as the 3.5 inch FDD became mainstream, and Laptops took over in the portable segment of the market. The same cascading of technology from the High End and Main series continued with both the Laptop and Notebook series. It can also be observed that the number of new model introductions is about two per year across the entire family. This family was only one of several families of personal computers introduced by NEC (Methé, Toyama, and Miyabe, 1995). Although serving various market segments these various families were related to one another through the compatibility of operating system and microprocessor. Consequently, in a given year, NEC may have introduced

Figure 3.2
NEC PC-9800 Product Family Group

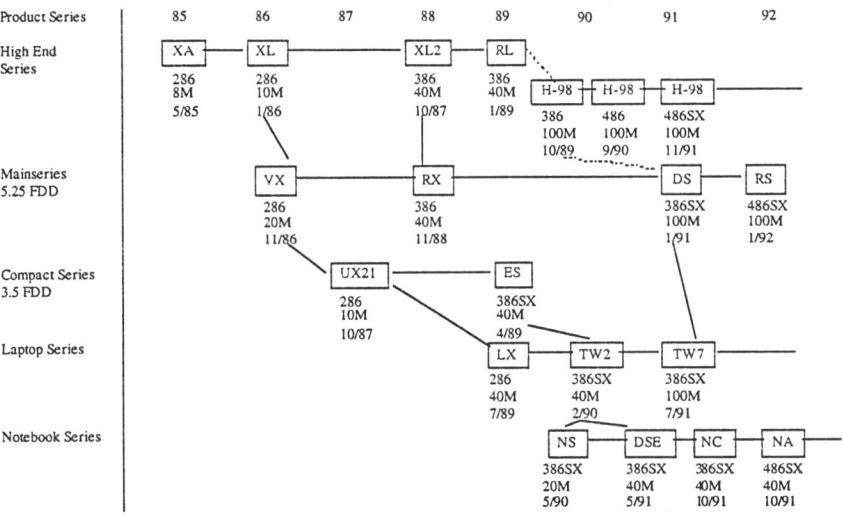

more than two new models per family group, along with variants of each model.

In order to support the dominant technical logic, which can be classified as a "product innovation treadmill," the structure of the research activity required a close connection between those actually making the product and those designing and developing the product. As a result, laboratories are often colocated on the same grounds as factories. This close proximity of designer and manufacturer facilitated the sharing of technical information and shortened the product development cycle. Although this colocation is quite common, one of the most noted is the Toshiba works at Ome, which was responsible for developing that company's laptop computer.

Not only is the physical location of the research facility a factor, but other organizational routines were structured to help keep this system going. Although the standard research and development routines of Japanese companies vary from company to company, these practices emphasize efficiency, both with respect to human and material resources and especially to time. One of the first steps a Japanese engineer takes when handed an assignment is to study the relevant patent database to determine if a useful answer already exists. This also helps to explain why many research laboratories in Japan have their own patent library on site, even with space at a premium. This approach is predicated on the assumption that the engineer is dealing with a known technology trajectory, whether it was Basic Oxygen Furnace (BOF) in steel processing or Metal Oxide Semiconductor (MOS) in integrated circuit processing. Efficiency in moving along each trajectory is key in the research process. This also helps to explain why engineers will be willing to move from the research laboratory to the factory floor. Not only does this support the dominant technical logic, but this behavior is rewarded by the organization in that research success is defined in terms of market success. Engineers in this kind of environment are rewarded more for moving an idea from concept to product than from concept to publication. The drive for efficiency, especially in the use of time for speeding a product to market, tends to focus an engineerís attention on finding an answer rather than allowing the luxury of pursuing interesting questions. These interesting questions may open up new avenues of research, which in turn raise even more questions. Consequently, "question raising" or exploratory research tends to be crowded out by "question-answering research." As a result, creativity tends to be channeled along defined pathways.

TECHNOLOGICAL CHALLENGES IN THE ELECTRONICS INDUSTRY

The pace of technological change is presenting a tremendous challenge to firms operating in the electronics industry. At the same time, the demand for greater variety of electronics products is leveling off or falling (Table 3.1). The markets that Japanese companies sell in are becoming increasingly saturated. Also, the consumer is becoming more sophisticated and not as easily convinced that a VCR, PC, or TV with new buttons or knobs is a "must buy." At the same time the knowledge base that was exploited in the past to develop new products is shrinking, creating a technological challenge for Japanese companies. This technological challenge is occurring on two fronts: increasing miniaturization of components and greater systematization of products. At the component level, the impact of increasing miniaturization is especially evident in integrated circuit development. The progression of technological innovation into the 256M DRAM and beyond is pushing the technology horizon for design and fabrication of integrated circuits past the solid state physics knowledge base that has supported all previous integrated circuit innovation and into quantum physics. The uncertainty of when, and possibly if, the shift from solid state physics to quantum physics will occur has generated a sense of urgency among integrated circuit manufacturers and the upstream firms that supply them in planning for new chip development.

This sense of urgency is further heightened by the trend toward greater systematization. The boundaries between what is a component and what is a system are blurring at an accelerating pace. This process began with the first microprocessor and is hitting its stride with the emergence of the new sixty-four bit microprocessors. The impact of this process is being felt most strongly in the downstream industries that use these microprocessors. The emergence of the multimedia field is a result of these innovations.

Even if multimedia did not exist on the horizon, the higher degree of systematization would be fed from another source, the increasing interdependence of innovation in the hardware and software areas of electronics. The ability to sell electronic products to an increasingly sophisticated and satiated consumer depends more on the ease of use of the product, not only as a stand alone, but also as part of a broader network of products in which it can interact. To provide such capabilities, more powerful systems software and object technology (object-oriented operating systems) need to be developed (Boyd, 1993). These new software innovations will require even greater sophistication in both the memory and microprocessor technologies to make them work. Consequently, the need to produce even more systemic products will further push components into miniaturization.

JAPANESE RESPONSE TO TECHNOLOGY CHALLENGES

This positive feedback cycle between systemization and miniaturization is beginning to lead electronic companies into uncharted technological waters. Not only is the underlying knowledge base for integrated circuit production shifting, but the skills needed to design and produce the networked, user-friendly multimedia products of the future have pushed electronics companies into deeper explorations of such fundamental areas of science as the nature of intelligence, language, and learning. These technological waters are uncharted in the United States as well as Japan. Couple these facts with the pace of technological catch up by Japanese companies, which has already occurred, and the result is that many Japanese companies no longer are confronted with the comfortable choice of "make or buy" the technology. The knowledge upon which twenty-first century electronics products will be based is now being developed, and Japanese companies are aware that they must be a part of the process that "makes" or generates that knowledge or be left behind in the competition for emerging markets. The rise in the number of "basic research" institutes in Japan is indicative of Japanese companies' strategic understanding of the situation (Table 3.2).

The number of basic research institutes in Japan increased steadily throughout the 1980s. Not all companies established a basic research institute in response to the environmental challenge to the dominant logic, however. And those that did, did not do so in an uniform manner. The remainder of this

Table 3.2
Number of Basic Research Institutes Established in Japan

Year	Number
1976	2
1977	1
1978	1
1979	1
1980	1
1981	4
1982	1
1983	6
1984	7
1985	6
1986	5
1987	6
1988	13
1989	11

Source: Chogin Sogo Kenkyu Jo, 1991.

chapter explores some of the differences in how Japanese companies in the electronics industry responded to the need for greater creativity.

BASIC RESEARCH AT JAPANESE ELECTRONIC COMPANIES

The various approaches taken toward developing the knowledge exploration capabilities of ten leading Japanese electronic companies are described in terms of several organizational variables: the mission that is ascribed to the research organization by the top management of the company, that is, what is the value orientation of the new technical logic; the structure for carrying out the new mission, that is, is a dedicated location provided for housing the knowledge creation activities or are these distributed throughout the company; the systems used to select and evaluate both researchers and research project, that is, are new procedures designed to fit the new activity or are they the same procedures from the knowledge exploitation activity used, and the people selected to carry out the knowledge creation activity, that is, how socialized are these into the older dominant technical logic (Table 3.3).

In examining the companies, several will be selected as representative of the type of approach being taken to establish a knowledge creation technical logic. These will be NTT, Hitachi, NEC, and Toshiba. Examples will be drawn from other companies when they help to illustrate a point. NTT is in a unique position among the companies arrayed in that it does not make any of the

Table 3.3
Various Approaches Taken toward Developing the Knowledge Exploration
Capabilities of Ten Leading Japanese Electronic Companies

	Mission	Structure	Systems	People
NTT	science, new knowledge 20+ years	Dedicated 100% corporate 1985 1991	Evaluation and selection geared to BR	BRL ~ 180 CSL ~ 200 Cross & new hire foreign researchers
Hitachi	science base of new business 10-20 years	dedicated 100% corporate New location 1985	Evaluation and selection modified for BR	ARL ~ 150 new hire foreign researchers
NEC	fundamental knowledge in new business 5-10 years	dedicated 85-90% corporate New location 1982-1989	Evaluation and selection modified for BR	Tsukuba ~ 100 cross new & hire foreign researchers
Toshiba	foundation for new business 5+ years	hybrid 100% corporate ARI w/in CRL 1991	Evaluation and selection essentially the same	ARI ~ 50 Cross & New Hire
Melco	foundation for new business 5+ years	hybrid 80-90% corporate ARF w/in CRL	Evaluation and selection essentially the same	~ 50 cross & new hire
Fujitsu	foundation for new business 3-5 years	Distributed basic research	Various programs to enhance creativity	About 1500 researchers at 3 Labs
Canon	foundation for new business 3-5 years	Distributed basic research	Various programs to enhance creativity	About 200 researchers at CRL
Sharp	foundation for new business 3-5 years	Distributed basic research	Various programs to enhance creativity	About 800 researchers at CRL and TRL
SEI	foundation for new business 3-5 years	Distributed basic research	Various programs to enhance creativity	About 576 researchers at Osaka and Itami
Sony	foundation for new business 3-5 years	Distributed basic research	Various programs to enhance creativity	About 500 researchers at CRL and research center

components that go into its telecommunication systems. It is essentially a service provider, having a local monopoly over local calls in Japan. It is facing competition in the long distance market as well as emerging telecommunication areas such as wireless communication. There are about 4,000 researchers in the organization, which employees about 8,600 people. There are twelve laboratories and two development centers. When NTT was privatized in 1985, a major reorganization of its R&D structure took place and the Basic Research Laboratory (BRL) was established. Currently there are about 180 researchers at the BRL. Another laboratory with strong basic research activities, the Communication Science Laboratory (CSL), was established in 1991. The BRL focuses mainly on the hard sciences of physics, while CSL explores human communication processes. The funding for the BRL is 100 percent from corporate funds, with no strings attached. The researchers only constraint is that the research must have something to do with telecommunications technology. There are about fifty non-Japanese researchers at the various NTT laboratories, with about half at the BRL. Most of these foreign researchers are working on their postdoctorates in some field of physics. Within the NTT research organization itself, roughly 15 percent of the researchers have a Ph.D., about 10 percent acquire it while at NTT. About 58 percent initially have a Master's degree, 18 percent with a Bachelor's, and the remaining 19 percent with some other technical degree. Many of the researchers working at the BRL are cross hired in from other laboratories in the NTT R&D system, but many are hired out of college as well. The mission of the BRL and CSL is to develop new knowledge. The only output expected is papers and presentations at conferences related to the field of research. Patents are also important as an output, but no work is done on commercializing the patent. The basic research activities at NTT are motivated by pure knowledge exploration and creation. Since NTT has always had a strong research ethic, the establishment of knowledge exploration activities is easier than in the other companies we will explore. NTT sets a baseline from which to compare the endeavors of the other companies.

Hitachi is representative of the sogo denki companies that compete in the electronics industry in Japan. Hitachi also occupies a unique situation among the companies establishing knowledge creation activities. It has a strong knowledge exploitation logic. This is illustrated by the extensive R&D system within the company. There are nine corporate laboratories and twenty-six divisional laboratories in this large, integrated electronics manufacturer, with four thousand three hundred people involved in research at the corporate laboratories and eight thousand six hundred people involved at the divisional and works laboratories. However, of the nine corporate laboratories, eight are closely tied into the day-to-day activities of the divisions. About 70 percent of their research budget is from contract research from the divisions and works. The other 30 basic research Various programs to enhance creativity About 500 researchers at CRL and research center percent comes from corporate coffers. The Central Research Laboratory, which in many ways acts as a hub for the other corporate laboratories, receives about 50 percent of its research funds from the corporation and the remaining 50 percent from the divisions and works.

Top management at Hitachi felt they needed to establish a new set of organizational routines for conducting exploratory research. As a result, the company decided to establish a new laboratory, the Advanced Research Laboratory (ARL), in 1985. In doing so, the company decided that 100 percent of the ARL's funding would come from corporate headquarters. Consequently, it is the only laboratory in Hitachi's research system that is structured to be totally independent from the day-to-day activities of the operating units. It is also physically quite removed from the rest of the company's laboratories and factories. The ARL was originally placed in Hitachi's research campus, which houses the Central Research Laboratory (CRL) and other related facilities. This was initially done more out of space considerations and the fact that the new laboratory's building and equipment had not yet been completed. The ARL was moved from the original site at the company's research campus to a physical location removed from the rest of the company's activities. Considerations such as space, cost of land, and the like were important. However, the thinking that by placing the ARL away from other research laboratories meant that researchers could truly focus on their question-raising research and not be caught by sociometric forces drawing them into question-answering research was given weight by the decision-makers. There are currently about 150 researchers at the ARL, of whom about 4 are non-Japanese. In time the number of researchers should expand to 200. There are also further hopes to internationalize the research agenda and to increase the number of non-Japanese researchers to about 10. In addition to conferences, contact with the outside is done through publishing papers and joint research. One of the ARL senior researchers is currently participating in an ERATO project for example (Methé, 1995). In addition, some ARL researchers may leave for other laboratories in the company or may exit the company completely. Some have gone into academic institutions, but none have entered other companies. Further, the choice of projects plays a role in controlling the degree of connection to the rest of the company.

The ARL is dedicated to the long-range, high-risk, new technological knowledge-generating projects that make up basic research in Japan. It was stated that the ARL belongs to the company and is not a university laboratory. As such, the type of projects suggested and selected should have some kind of product or process application to the company. What distinguishes the ARL from other laboratories is the time frame and degree, both in terms of difficulty and number of knowledge-generating activities involved. All projects may have a ten to twenty year, or longer time frame, before the knowledge generated is expected to be converted into products. These projects are expected to result in papers and patents, in that order. Products were not expected to be a direct result of research, although other laboratories were expected to use the knowledge gained to develop products. In conducting some research in electron physics, it was necessary to develop new tools, one of which, a holography electron microscope, could result in a salable product, it was thought.

Although not as large as Hitachi, NEC is also a sogo denki company. There are nine laboratories and three development laboratories, which are corporate laboratories, within the R&D Group. There are ten operating product/technology groups, such as the Semiconductor Group or Personal C&C Group, which are made up of a number of divisions, each with laboratories. There are about one thousand six hundred employees in the R&D Group. It receives about 1 percent of total sales as its budget. Of that, 10 percent is devoted to basic research. The Divisions receive about 10 percent of sales as their R&D budget. NEC has also established a basic research facility within the R&D Group as a brand new laboratory away from other laboratories. NEC had a basic research laboratory as early as 1982. This was located in the Central Research Laboratory complex. In 1989 NEC established a new laboratory complex at Tsukuba. Although the new complex was established to conduct knowledge exploration activities, it has included several applied research labs inside the new "basic research" laboratory. Further, the funding for the laboratory is not totally from corporate sources, it also receives some of its funds, about 10 to 15 percent from the divisions, as itaku kenkyu or "contract research." In this case, the new basic research laboratory is separated, physically, but not in terms of funding. As such it is not totally dedicated to basic research, in that some facilities and researchers have as their explicit charge the development of products or work on processes. On average for the entire R&D Group, this type of research makes up 30 percent of the research budget. There are currently about fifty researchers at the Tsukuba laboratories, with plans to go to about one-hundred. There are also about ten to fifteen foreign researchers at the laboratory. The selection and evaluation process has been modified to account for the longer time frame involved in doing knowledge exploration type of research, but the process is similar to that at other laboratories. Patents and papers are considered important in the evaluation. The researchers are cross hired from other NEC laboratories as well as brought in from university. About 20 percent are Ph.D.'s in the current group of researchers at the Tsukuba Laboratories. The mission differs slightly from that of Hitachís, in that it is tied more closely to what are considered core technologies for businesses that NEC is in or may enter in the future.

Toshiba is also a sogo denki company in Japan. Toshibás R&D organization underwent a large reorganization in 1992. The twelve research institutes, as the laboratories were called, were reconfigured along technological lines into five research institutes. These along with the ULSI Research Institute and the System Software Production Technology Research Institute were all brought under one organization, the R&D Center. Toshiba also has a large number of divisions and laboratories at these. Currently, some twelve-thousand employees are said to be involved with some kind of R&D activity within Toshiba. There are about one thousand and six hundred researchers at the R&D Center. At the time of the reorganization a Basic Research Institute (BRI) was established within this R&D Center. The R&D Center receives about 53 percent of its funding from corporate and about 43 percent from contract research from the divisions. The remainder is contract research from outside organizations such as the government. The BRI receives all of its funding from corporate sources. Total R&D budget devoted to knowledge exploration type basic research is 10 percent, however, the BRI only gets about 2 to 3 percent. The remainder is carried out in other research institutes. There are about fifty researchers within the BRI. These have been cross hired in or brought in as new hires. Although in the R&D Center as a whole there are about 15 percent Ph.D's., 65 percent Master's and about 20 percent Bachelor's, the BRI has about thirty Ph.D's, or about 60 percent of its total. The mission of the BRI, as with most of the knowledge exploration activities in Toshiba, is building a strong foundation for new business activities, with a time horizon of more than five years.

Fujitsu is representative of the next group of companies. These companies have not chosen to establish a dedicated organizational structure to house their knowledge exploration activities, but have instead decided to utilize various programs to stimulate knowledge exploration activities. There are three laboratories within Fujitsu Laboratories, which became a wholly owned subsidiary of Fujitsu Corporation in 1968. Each laboratory is responsible for some technology/product area that Fujitsu is involved in. For example, the Kawasaki laboratories are concerned with computers and telecommunication systems, whereas the laboratory in Atsugi is concerned with components such as semiconductors. The Information Social Science Laboratory was established to examine such questions as these: Where and how will the information society progress? What kind of "tools" will people use in the twenty-first century? How do Japanese people recognize kanji, and how is the pattern physiologically processed and stored? Also, some "hard" science is being done in this laboratory concerning fields such as biosensors. In terms of its R&D budget, about 60 percent comes from corporate, with the other 40 percent coming from the divisions as contract research. About half of the corporate money is tied to supporting current business activities, leaving about 20 percent for longer term knowledge creation type research activities. This R&D budget of Fujitsu Laboratories makes up about 12 percent of the total spent on R&D, with the balance being spent in the division's laboratories. As noted, no physical basic research laboratory was established at all. In its place a number of systems that would allow researchers to receive resources to follow difficult and challenging research projects that would possibly result in products or processes for the company were put in place. One of the systems was called the *My Way Project*. Begun in 1991, it is open to all 1,150 researchers within Fujitsu Laboratories, and allows five to six researchers to receive awards that allow them to work on their own projects fully funded for up to three years.

PATH DYNAMICS OF BASIC RESEARCH DEVELOPMENT AT JAPANESE ELECTRONICS COMPANIES

There are some important aspects of the development of knowledge exploration activities within Japanese electronics companies that are not readily apparent. Each company that established a dedicated laboratory for housing these activities arrived at this through a different set of path dynamics. Before describing these, it is important to note that the common element among all company attempts was to somehow insulate the knowledge creation activities from the dominant technical logic of knowledge exploitation.

This appears in the case of Mitsubishi Electric (MELCO). The basic research laboratory appears to have evolved in the central research laboratory, as more and more corporate laboratories dedicated to specific technology areas were established. Some area of long term interest to the company would be worked on inside the CRL. As technological knowledge accumulated and was being converted into products, a critical mass of researchers would be developed and this group would "spin out" into a new laboratory. These dedicated technology laboratories appear to take on more of the longer term product-development activities that were once done by the central research laboratory. As a result, the central research laboratory has more freedom to pursue even broader ranging and further time frame technology projects.

Hitachi represents another path. Hitachi has set up laboratories in the past with the idea that these would do knowledge exploration activities. The current Central Research Laboratory located at the Kokubunji campus was originally set up to do basic research. It was gradually captured by the pull of the divisions. It still gets about 50 percent of its budget from corporate sources and devotes these resources to long term knowledge exploration. As noted above, this was one of the motivations for setting up the ARL and for physically and financially separating it from the rest of the R&D system. This path sounds similar to that followed by NEC, except in this case NEC consciously choose not to totally separate the laboratories doing knowledge creation activities from those doing knowledge exploitation activities.

EVALUATION AND CONCLUSION

What is important to recognize about these various paths is that they are not randomly taken. As noted, each of these companies had to find a way to insulate the new knowledge creation logic from the dominant knowledge exploitation logic. The new logic is always in danger of suffering from an extreme form of "Greshem's Law." In order to prevent this, barriers have to be developed in order to insulate the new routines. These barriers can be physical, organizational, disciplined based, or personal (Morton, 1967; Lane, Beddows, and Lawrence, 1981). If the laboratory is physically remote it is less likely to be influenced by other parts of the organization. Likewise, if it receives all its funding from one source, it is free from influence from other sources. By grouping researchers according to their disciplines, the mind set and specialized vocabulary sets up walls that other disciplines cannot breach. Also, the very act of calling an activity "basic" research sets up a disciplinary barrier in terms of the mind set for carrying out the research process. Finally, by hiring in only new people who have not experienced the other areas of the organization, it is possible to socialize them to ignore other parts of the organization. The creation of these barriers cuts down communication and slows the cross-learning between the various groups within the organization. If the object was to simply insulate the new routines form the old, the process of creating barriers would be enough. However, each of these companies must compete in future as well as current markets. The newly created technological knowledge must be transferred to some current or future operating division if the company is to avoid "fumbling the future." Consequently, some type of bond must also be created to link the knowledge creation activity with the knowledge exploitation activity (Morton, 1967; Lane, Beddows, and Lawrence, 1981). Again, the bonds can be created by colocation, as NEC did by putting basic and applied activities together in the same laboratory, organizational as MELCO does by funding some of its basic research through divisional contract research, disciplinary as with Toshiba's combining of research institutes along common technological lines such as materials or energy, or personal, by cross-hiring researchers from other laboratories in the system.

The process by which each of these companies attempts to balance the barrier and bonds building can be compared and contrasted (Figure 3.3). Hitachi is close to totally insulating the knowledge exploration activity from the knowledge exploitation dominant logic of the divisions. On the other side, Fujitsu is representative of those companies that have more bonds than barriers. As a result, the knowledge exploration routines are much more integrated into the rest of the organization and more open to the influence of the dominant technical logic. About in the middle are companies like Toshiba and Mitsubishi Electric.

**Figure 3.3
Integrated versus Insulated Spectrum**

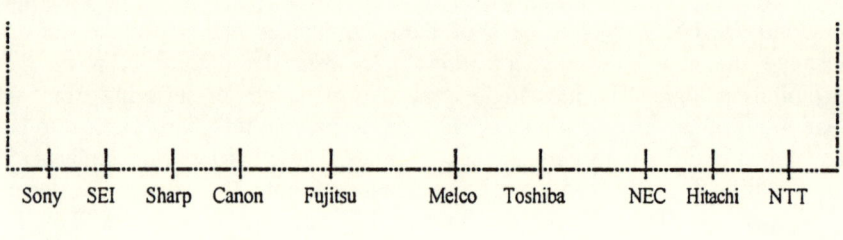

Because each of these companies has only recently attempted to build up its knowledge creation routine it is too soon to evaluate which of the various strategies of mixing barriers and bonds will yield the best balance of knowledge exploration and exploitation. It is interesting to note, however, that the previous experience of each of the companies appears to have something to do with the current approach. Although the size of the companies also appears to play a role, the initial conditions that set the company on the path to establishing a basic research laboratory also appears to play a role. It is also important to note that within the Japanese economy, other institutions that carry out the knowledge creation routines, such as universities, are weak.

The often-voiced laments by the Japanese about a "Nobel Prize gap" was one manifestation of this sensitivity. Several Japanese authors and researchers also noted the need for greater "creativity" on the part of Japanese companies if they are going to be able to contribute to the future of humankind, let alone survive in the competition of the future (Murakami & Nishiwaki, 1991). Even in interviews conducted for this report over the past two years, Japanese researchers have seriously raised the issues of whether they are creative or even capable of creativity in the sense of a truly ground-breaking invention or discovery. Companies feel that they are the only institutions in Japanese society able and willing to carry out the knowledge creation routine. If this is so, and Japanese companies are the only institutions creating new knowledge in Japan, there is a danger of future international disputes over intellectual property rights (Methé, 1995).

A more likely scenario is that a combination of recession and the strength of the dominant technical logic will cause these new routines to be overwhelmed by the old. This would cause the research institutes to fail and be absorbed back into the organization. This would also have international consequences. Such a failure would result in the atrophication of the capabilities needed by Japan to move away from being an importer and exploiter of knowledge from the West. The current round of trade friction is a symptom of this unhealthy situation.

The likelihood of this scenario places even greater pressure on the other institutions in Japan's economy to take on the knowledge task. This is especially true for Japanese universities. In order for Japan to truly move into the ranks of a leading nation, it must bear the burden of creating new knowledge. By not developing these capabilities it is likely that Japan will be consigned to the position of follower. Although this may at first glance appeal to American companies, what it really means is more of the same. For those who would worry about a Japan capable of creating and utilizing knowledge, it should be remembered that there are more than enough technological, economic, and social challenges to go around. Markets will open up to those who are creative enough, brave enough, and resourceful enough to accept those challenges.

REFERENCES

Abegglen, J. C., and Stalk. Kaisha. *The Japanese Corporation.* New York: Basic Books, 1985.

Arrison, T. S., Graham Bergsten, and Harris. *Japan's Growing Technological Capability: Implications for the U.S. Economy.* Washington, D.C.: National Academy Press, 1992.

Betz, F. *Managing Technology: Competing Through New Ventures, Innovation Corporate Research.* Englewood Cliffs, N.J.: Prentice Hall, 1987.

Boyd, J. "The Computer World According to Sun." *The Japan Times,* 19 July 1993, p. 19.

Chogin, S. K. *R&D katsudo no guroobarizeshun: Kokunai kigyo kara sekai kigyo e no michi* [R&D globalization activities: The path from a country-based industry to a world-based industry]. Tokyo, Japan: Nihon Choki Shiny Ginko, 1991.

Choy, J. "Japan's Computer Software Market: Where Are the Japanese?" JEI Report. Tokyo, Japan: Japan Economic Institute, No. 13A, April 1, 1994.

Fruin, M. Organizational Architecture and Innovation: Knowledge Works at Toshiba. Ann Arbor, Mich.: University of Michigan, November 13, 1992.

Gibbs, W. W. "Basic strategies: Japanese Companies Cultivate Research Labs Sown in the U.S." Scientific American 270, no. 4 (April 1994): 114–15.

Hicks, D., and Hirooka. "Defining Basic Research in Japanese Companies and Science in Japanese Companies: A Preliminary Analysis." Tokyo, Japan: National Institute of Science and Technology Policy, September 1991.

Hirano, Y., and Nishigata. "Basic Research in Major Companies of Japan." Tokyo, Japan: National Institute of Science and Technology Policy, January 1990.

Ishii, Y. "A Neglect of Research May Doom Japan's Future." Economic Eye, 9–11, Vol. 14, No. 1, Spring 1993.

JICST. Kigyo kenkyu kaihatju joho mappu [Company research and development information map]. Tokyo, Japan: Nihon Kagaku Gijutsu Joho Center, 1992.

Lane, H., W., Beddows, and Lawrence. *Managing Large Research and Development Programs.* Albany, N.Y.: State University of New York Press, 1981.

Management and Coordination Agency. Linked Input-Output Tables. Tokyo, Japan, 1985.

———. Linked Input-Output Tables. Tokyo, Japan, 1980.

———. Linked Input-Output Tables. Tokyo, Japan, 1975.

Methé, D. T. *Basic Research in Japanese Electronic Companies: An Attempt at Establishing New Organizational Routines.* In *Managing Technology Through Organization: U.S. and Japan Approaches,* edited by J. Like, J. Ettlie, and J. Campbell. New York: Oxford University Press, 1995.

———. "Techno-Global Competition in the Integrated Circuit Industry." *Advances in Applied Business Strategy,* vol. 3. JAI Press, 1992.

———. *Technological Competition in Global Industries: Marketing and Planning Strategies for American Industry.* Westport, Conn.: Quorum, 1991.

———. "Technology, Transaction Costs, and the Diffusion of Innovation: The Evolution of the United States and Japanese DRAM Integrated Circuit Industries." Unpublished Ph.D. Dissertation, University of California, Irvine, 1985.

Methé, D. T., Toyama, and Miyabe. The End of the NEC Shogunate: A Historical Overview of Standards, Technological Innovation and Strategy of Japanese Personal Computer Firms.î University of Michigan Working Paper, 1995.

Morton, J. A. "A Systems Approach to the Innovation Process." *Business Horizons* Summer 1967.

Murakami, T., and Nishiwaki. *Strategy for Creation.* Cambridge, England: Woodhead Publishing Limited, 1991.

Muto, E., and Hirano. "Government Laboratories and Basic Research: Towards the Promotion of Basic Research in Government Laboratories." Tokyo, Japan: National Institute of Science and Technology Policy, September 1991.

Odagiri, H., and Goto. "The Japanese System of Innovation: Past, Present and Future." In *National Innovation Systems: A Corporative Analysis,* edited by R. R. Nelson. New York: Oxford University Press, 1993.

Oka, H. "The Industrial's Sector's Expectations of Engineering Education." Economic Eye 14, no. 1, (spring 1993): 12–15.

Prahallad, C.K., and Bettis. "The Dominant Logic: A New Linkage Between Diversity and Performance." *Strategic Management Journal* 7 (1986): 485–501.

Shapira, P. "Lessons from Japan: Helping Small Manufacturers." Issues in Science and Technology, 8, no. 3 (spring 1992): 66–72.

Shōhisha ya ryōhaten o ki ni shitsutsu kaden meka wa ugoku. "'Home Appliance Makers' Movements Focus on Consumers and Retail Stores." Tokyo, Japan: Nikkei Electronics, No. 558. July 6, 1992, pp. 120–41.

Tushman, M. L., Newman, and Romanelli. "Convergence and Upheaval: Managing the Unsteady Pace of Organizational Evolution." California Management Review 29, no. 1 (fall 1986): 29–44.

Weber, H. University-Industry Cooperative Research Centers in National Universities in Japan. Report Memorandum #93–7. Tokyo, Japan: Tokyo Office of U.S. National Science Foundation, 1993.

Part II

THE JAPANESE TECHNICAL STANDARDS SYSTEM AND TECHNO-COMPETITION IN SELECT JAPANESE INDUSTRIES

4

Globalization and the Role of Standards

Koji Tanabe

OUTLINE OF JAPANESE INDUSTRIAL STANDARDS

In Japan, Japanese Industrial Standards (JIS) have been established as national voluntary standards since 1949. JIS covers 8,018 standards in the industrial and mining fields. To develop JIS standards, it is essential to ensure transparency in the process of standards development and to get the consensus of interested parties such as manufacturers, users and consumers, and academic societies. Drafts of JIS standards, which are usually prepared by the relevant industrial associations or academic societies, are deliberated by the Japanese Industrial Standards Committee (JISC), which includes representatives of the interested parties. JISC is the national standards body and a member of the International Standards Organization (ISO) and the International Electrotechnical Committee (IEC).

As a voluntary certification system, the JIS Marking system has played an important role in promoting the dissemination of JIS, product quality improvement, and in providing information on product quality for consumers. Today 15,367 factories, including 280 overseas factories, have JIS Mark certification on 785 products. Sixteen foreign inspection bodies, including the Underwriters Laboratory (UL), the American Bureau of Shipping (ABS) Industrial Verification, and the United States Technical Committee (USTC), have been designated to promote the JIS Marking system to factories in the United States and in other foreign countries.

THE SIGNIFICANCE OF STANDARDIZATION

The president of a famous camera maker once described the essence of standardization as follows: A Japanese visiting the United States buys a camera there, loads it with film he bought in Canada, and develops the film in Britain. He does this as a matter of course without any problem. Without standards, none of this would be possible. Moreover, what makes this series of activities possible is that conformity assessment systems that test and certify the conformity of products with standards are well established throughout the world.

On the other hand, when we travel abroad and bring with us electronics such as a PC or a razor, the plug socket and voltage vary from place to place, making it inconvenient to use these products. These inconveniences are due to a lack of successfully harmonized standards. In our global economic society, these types of problems occur daily, in our personal activities as well as in production activities, when various items are not internationally standardized.

Standardization, the establishment of global and uniform rules concerning technical items, aims at creating utility for society at large by setting rules for technical specifications of products such as performance and safety level. To a large extent, standards demonstrate the characteristics of public goods whose benefits are available to the whole society. It is the progress of industrialization, throughout the nineteenth century, which caused the spread of standardization to production sites at factories and into everyday life. Today, standardization is essential in a modern industrialized economy.

Standardization not only determines product standards, it also includes production systems and process standards, such as the ISO 9000 series, which sets quality management system standards.

PURPOSES OF STANDARDIZATION

1. Securing of Interoperability and Coordination of Diversity (Simplification): To enable the replacement and interconnectability of parts, and to facilitate mass production and cost reduction through the elimination of unnecessary diversity, such as the securing of interface of bolts and nuts, light bulb sizes, floppy disks (3.5 inch), merchandise bar codes, A-type paper, dry cells, telecommunications network protocol, etc.

2. Promotion of Mutual Understanding: To promote understanding of manufacturing and design, business transactions, and other facets of production and products, by mutual decision making. Examples include terminology, drawings, symbols, logos, musical scores, language, and other items that facilitate communication, as well as test methods that serve as objective criteria or assessment of standards.

3. Clarification of Quality: To eliminate goods of inferior quality and to provide criteria for purchasing by designating levels of product quality, such as performance characteristics and providing information on labels. Examples

include air conditioners (cooling capacity), instantaneous water heaters, (hot water supply capacity), and critical point testing method for shape-retaining alloys.

4. Safety, Health, and Environmental Protection: To enhance social benefits by protecting consumers and setting forth requirement standards for social and environmental protection. Examples include seat belts (safety), wheelchairs (personal welfare), slag aggregate (recycling), emergency exit marks and public guide maps, safety signs such as "Danger," industrial wastewater test methods, etc.

GENERAL TRENDS IN STANDARDS AND ISSUES FACED BY PUBLIC ORGANIZATIONS

Globalization

Markets are increasingly becoming globalized, while production and distribution systems are being organized beyond national borders. Globalization is drastically transforming the structure of corporate transactions and the system of each country's economic activities. In short, a borderless economic society is emerging throughout the world.

In such an international community, the standards and conformity assessment system of each country is vital to the international trade of goods and services. There have been many cases where international business transactions did not go smoothly due to conflicting standards between two trading countries.

Regarding the results of conformity tests, with government-established technical regulations, there are instances when one country does not accept the findings obtained by another country's testing laboratories. In such cases, product samples need to be submitted to the other country and tested there, thus requiring a lot of time and money. Accordingly, systems pertaining to standards have often constituted a major barrier to the international trade of goods and services. In response, certain international efforts have been made to eliminate this barrier.

First, the activities of ISO/IEC (the International Standards Organization and the International Electrotechnical Committee) should be mentioned. Each country has government-authorized agencies or groups that prepare or issue national standards. A country's national standards are determined through the consensus of the parties concerned. The ISO, established in 1947, is a gathering of standard development bodies, and various countries with common interests have cooperated together to reach agreements through such forums as ISO and IEC. International standards have been decided upon and applied in various nations. The ISO and IEC serve as advisory bodies to the United

Nations and pursue activities in conjunction with more than three hundred international organizations, including the World Trade Organization (WTO), the United Nations Environment Programme (UNEP), and the International Labour Organization (ILO).

The relationship with the WTO is especially critical, as shown by continuing efforts—the WTO's Agreement on Technical Barriers to Trade (TBT). At the 1980 Tokyo Round of the General Agreement on Tariffs and Trade (GATT), the TBT Agreement was concluded and rules were established in accordance with the following principal objectives: (i) each country shall adopt ISO/IEC standards as much as it is possible to do so; (ii) the conformity assessment system of each country shall be opened to foreign countries and they shall be treated without discrimination; and (iii) when standards and certification systems are to be established or amended, transparency shall be ensured by giving the WTO prior notice and by following proper international procedures.

At the Uruguay Round, it was decided that the TBT Agreement would cover voluntary standards as well. Under the TBT Agreement, however, there is an escape clause, which stipulates that there is no obligation to make certain items consistent with international standards; these include items stemming from climate and natural features, and items concerning health, safety, and the environment, among others. Since each country's items pertaining to health, safety, and the environment often take the form of technical regulations, it is not easy to harmonize the standards. Efforts are currently being made to minimize barriers to trade by acknowledging each other's certification systems.

Standards for actual intercompany commercial transactions and for the goods and services of individual companies often contain specifications that are more stringent, in terms of quality, than those contained in standards coordinated internationally. Thus, there is little incentive for a company to make national standards compatible with internationally established ones. Even though the standards themselves vary, the rules of the WTO and the spirit of the ISO/IEC are international in nature and aim at making the public control of goods and services possible through unified world markets. As such, attempts to narrow the gap between social interests and the interests of individual companies are ineffective unless coordinating activities are carried out by public organizations.

The international coordination of standards is time consuming because it attempts to change systems that have been established based on the history and social culture of each country. For this reason, the reduction of conformance costs through mutual recognition among countries is a realistic approach. Yet, since conformity assessment systems and past experiences—as well as the social values and customs prevailing in each country—vary for different nations, it is often difficult to approve each other's certification. Thus, it is critical to increase widespread confidence in relevant organizations. The coordination of public organizations is indispensable to such an approach.

It is important to create a harmonization program between countries by taking the building-block approach, starting with areas in which efforts to heighten mutual confidence can be undertaken with ease, and then gradually cover other more difficult areas.

Accordingly, as the international harmonization of standardization systems becomes increasingly important, each country's coordination capability and support for the involvement of the private sector in international standardization organizations will become exceedingly vital.

RAPID TECHNOLOGICAL CHANGE

Standards development in the areas of high technology, a field which is ever changing and advancing, is of great importance to many countries. Today, the speed of technological change is picking up in several areas, including the information technology sector. At the same time, the internationalization of markets is expediting the diffusion of technology from one country to another.

The standardization of goods and services has mainly focused on establishing technical rules of practice to facilitate mutually beneficial relations between companies, including improving the quality and interoperability of products while maintaining reasonable prices and stable supplies. While such standardization has been effective for technology that is not changing rapidly, de jure standardization of rapidly changing technology often falls behind schedule, and products of specific companies are commercialized under de facto standards. At times, the coexistence of several systems puts consumers at a major disadvantage and can result in social loss.

The preparation of standards by the ISO/IEC or by a national standardizing body is aimed at establishing rules of mutual control based on a consensus of the stakeholders involved. This implies that both users and producers participate in the process and can thereby maintain their common interests. But in fields where technological progress is being made at a fast pace, there are cases where standardization is undertaken as de facto standards, even if a currently emerging technology has no users as yet. In this case, since technical information is not properly disclosed, end users cannot properly compare it with other competing technologies. There are cases where a consortium of developers is formed to prepare de facto standards; however, there are also cases where high barriers to entry into consortia exist, and doubts about violations of antitrust law surface due to a tendency to monopolize technology.

Nevertheless, the details of de jure standards are clear and open, and their establishment/amendment process should be, and is, transparent. Membership is also open. However, standardization requires time, and frequently lags behind technological progress. More important, in fields where technological progress is advancing rapidly, the pursuit of standardization is tantamount to seeking changes within established rules of behavior. Consequently, this limits ensuing technological changes and may delay advances in technology.

Several attempts have been made in recent years to capitalize on the flaws and strengths of de facto and de jure standards. The first attempt pertains to the development of de facto standards. Prompted by the formation of international consortia at the research and development and demonstration stage, activities are being vigorously undertaken by organizations aiming at relatively open standardization. Whereas consortia were headed by one company in the past (the IBM consortium, for example), there are now cases like the Internet Society and DAVIS (a standardization society pertaining to video on demand) where several companies form a consortium under the council system. This can be deemed an intermediate form that brings de facto standards closer to de jure standards by creating more open standards and making procedures transparent. The line of demarcation between de facto standards and de jure standards, then, is becoming blurred.

The second attempt is the efforts of public organizations. While one such effort points in the direction of reducing time, another points in the direction of expediting the distribution of information.

The ISO and the IEC launched a system in 1994 under which private groups approved by them can make international proposals from the voting stage, like national standardizing bodies. Meanwhile, in the information sector, Publicly Available Standards were launched in 1995 that adopt, as a plan for international standards, items for which de facto standards have already been prepared. Thus, systems that reduce the amount of time required for preparing standards are already in place. Information concerning standards has been insufficient in the past. Nevertheless, efforts are being made to supply such information—required by users and industries to select the most effective standards—and to promote the formation of appropriate de facto standards, in the market, by developing systems that promptly distribute the latest information concerning standardization in the private sector.

The efforts of Europe's national standardizing bodies are similar in nature. They have introduced a "provisional standard system" that swiftly establishes standards that are in accordance with national standards as flexible and maneuverable de jure standards, even if perfect consensus is not obtained. The details of those standards are then disclosed. These bodies are aggressively incorporating the standardization efforts of the private sector and are stepping up international standardization efforts based on their own country's national standards. These initiatives are an attempt to explore new rules for sectors in which technology is making rapid progress.

IMPORTANCE OF INTERNATIONAL COORDINATION

Attaining economic globalization calls for the harmonization of national standards and conformity assessment systems. Standardization reduces procedures and costs and facilitates the exchange of information and consultation between the concerned parties. It is most desirable that the appropriate standardization of goods and services be effectively coordinated in the market through the process of supply and demand. Still, the system of each country has its own inertia, and these systems cannot be easily harmonized unless nations and public organizations consciously develop a coordination process. Moreover, in order to select the most effective standards, it is essential for all interested parties to obtain sufficient information. But since standards are an agreement designed to enable everyone to enjoy their benefits, information required to make proper decisions cannot be expected to be adequately provided solely through the market. This trend is especially conspicuous in sectors in which the pace of technological change is rapid. When standards are prepared after an agreement has been reached, in these sectors the number of interested parties increases and a greater length of time is required for such preparation since numerous difficulties emerge, until a new agreement is actually reached. Furthermore, the technology itself, a precondition of standards, may change and become obsolete by the time an agreement is reached.

Since the process of forming an agreement is a difficult one, it is possible that de facto standards that are deemed undesirable by users may be established. It is critical for public organizations to develop a system that expedites the distribution of information (for example, technology trend assessments), provided that such a system does not hinder the progress of technology. In addition, it is important to establish a system that allows de facto standards to be accepted as fair by as many people as possible, and that brings such standards closer to de jure standards.

International organizations such as the ISO/IEC or the standardizing bodies of many countries are mounting efforts in search of such a system. This type of system cannot be developed, nor can its operation be coordinated, through only the market mechanism. The role of governments and public organizations varies, according to the different socioeconomic systems in particular countries, but the crucial challenge of international coordination is to transcend the differences in each country's standardization system to develop a new, more global scheme in the standardization process. As a result of these new processes, a key issue is how to promote cooperation among the many organizations involved in the standardization process in the United States and Japan.

5

The Japanese Technology Infrastructure: Issues and Opportunities

John P. Stern

High atop Kasumigaseki, the heart of Japan's industrial bureaucracy, executives of the twenty major corporations that dominate Japan's electronics market are meeting. Officials of the Ministry of International Trade and Industry (MITI) busily record the discussion. The entire conversation is in Japanese. When the deliberations are over, Japanese business and government have reached a better understanding regarding the direction of new technologies.

This could be the opening scene in any number of paperback novels about business-government cooperation in Japan. But let me add some twists not mentioned in the typical pulp novel: representatives of foreign companies may have participated in the debate. The group may have discussed how to protect the intellectual property of a standard's leader. There may have been a serious disagreement concerning the proposed standard among several Japanese members of the group. And the discussion may have concerned the extent to which Japan should adopt, without change, a foreign standard rather than create its own unique standards.

Standardization in Japan is an immense topic, both in breadth and in detail. The Japan Industrial Standards (JIS) system, administered by MITI, contains more than eight thousand standards, covering everything from optical disks to heated toilet seats. Overseas, the JIS standards system is perhaps the best known Japanese standard's system, but other aspects of standardization affecting the electronics industry are under the jurisdiction of the Ministry of Posts and Telecommunications, the Ministry of Health and Welfare, the Ministry of Transport, or the Ministry of Construction.

Table 5.1
Who's Really On Top?

U.S. Leading Technology	Dependency on Japan
Semiconductors Steppers Dicing Saws Bonding Wire Ceramic Packages Epoxy for Plastic Packages TAB Equipment	High-Purity Silicon Wafers
Printers	Laser Diodes Laser Printer Engines
Computers	Floppy Disk Drives CD-ROM Drives Optical Disk Drives High-Resolution Displays Flat Panel Displays OCR Scanner Engines High-Performance Batteries
Multimedia	Video Conferencing Cameras CD-ROM Drives High-Resolution Displays Flat Panel Displays
Cellular Telephones	Gallium Arsenide High-Performance Batteries
"Smart Weapons"	Ceramic Packages Video Cameras Flat Panel Displays Micro-motors Gyrocompasses
Aviation	Gyrocompasses Flat Panel Displays Micro-motors Composite Materials Entertainment Systems
Satellites	Sensors Communications Module

But let us back up a minute. Before I say more about the Japanese technology infrastructure and technical standards in Japan, it might be useful to review why Japan remains an instructive case for the United States. After all, have we not been told that Japan is sinking under an overinflated yen and afflicted with backward technology. Is Japan not shrinking in influence on the U.S. economy as compared to the emerging markets of Asia?

What about the notion that the United States enjoys a huge technological lead over Japan, in all the right areas of information technology? When advocates of American technical superiority focus on particular products, they often point to semiconductors. It is true that, in 1994, the U.S. share of the world semiconductor market was larger than Japan's. But not by much: the U.S. share by revenue was 43.4 percent, while the Japanese share by revenue was 40.1 percent (World Semiconductor Trade Service,[1] 1994) less than a 4 percent lead. So, the United States is ahead by a nose. That is more encouraging than the 1990 ratio, when the Japanese semiconductor industry had more than a 10 percent lead. But that is nowhere near the 53 percent of the world market that the U.S. semiconductor industry enjoyed in 1984.

If we consider the "food chain" leading up to the information technology in which the United States excels, the situation is more disturbing (Table 5.1). As a result of technological dependency on Japan, the United States was unable to effectively punish a company that sold missile parts to Iran and perhaps Iraq; the U.S. semiconductor industry was panicked by a sudden shortage of epoxy resin for semiconductor packaging; and the U.S. computer industry has been unable at times to offer customers better or less expensive products.

If the United States is the world leader in information technology, at times it appears to be winning only by throwing everything overboard in its race to develop new technology. The next chart (Table 5.2) shows the information technology the United States buys from Japan, and what the United States ships to Japan. The United States has a surplus in computer CPU trade with Japan, but a deficit in computer parts. In many product areas, the United States exports virtually nothing: camcorders, facsimiles, film developing equipment, copier parts. In other areas, such as hard disk drives, American companies lead in design, but some of the same companies find it easier to buy the drives' original equipment as manufactured (OEM) in Japan. A few items of major Japanese surplus are baffling: Japan has racked up half a billion dollars a year, for a decade, exporting car stereos to the United States, but American companies have known how to make car audio equipment at least since Motorola's first product. So long as the United States imports more than $25 billion of high value-added electronics products from Japan year after year, Japan still matters.

Progress has been made in promoting U.S. sources of supply for certain critical technologies, but very little progress has been made in countering the growing ideological influence of Japan. Japan is an attractive model of

Table 5.2
1994 Japan–U.S. Electronics Trade Top Exports-Imports

U.S. Imports From Japan $ Billions	Product + % of Japanese Export Destined for U.S.	U.S. Exports to Japan $ Billions
6.29	Computer Parts (53.3%)	1.37
2.91	MOS IC Memory (50.1%)	0.52
0.34	CPUs (38.6%)	1.17
1.42	Hard Disk Drives (54.9%)	0.63
0.52	Telecommunications Equipment (32.3%)	0.26
0.82	Other MOS IC (27.0%)	0.29
2.69	Printers (49.5%)	
1.80	TV Cameras + Camcorders (43.2%)	
0.59	Computers w/ I/O Devices (37.2%)	
1.43	Photocopiers (48.0%)	
1.20	Copier Parts (39.6%)	
0.97	VCRs (28.1%)	
1.06	Displays (52.5%)	
0.91	Facsimiles (45.4%)	
0.82	Instrument Parts (45.4%)	0.55
0.67	CD Players (47.6%)	
0.63	Wired Telecommunications Parts (40.2%)	0.24
0.58	Car Stereos w/ Radio (47.8%)	
0.56	Microprocessors (20.9%)	1.13
0.52	Misc. Electronic Components (33.2%)	
0.42	Capacitors (18.3%)	
0.43	Switches (34.2%)	
0.35	IC Wafers + Dice (22.5%)	
0.55	Film Developing Chemicals (38.5%)	
0.48	Film Developing Equipment (31.8%)	
0.54	Batteries (30.5%)	
0.37	Blank Video Tape (44.6%)	
	Wireless Telecom Eq. Parts	0.27

Source: JETRO data, March 15, 1995

economic success for the rest of Asia. Let us remember that in 1960 Japan had a per capita GNP of only $477, barely 20 percent that of the United States. Fast forward to 1990, and Japan's per capita GNP exceeded that of the United States (Figure 5.1), and had multiplied nearly fifty-fold (Asahi Shimbun, June 14, 1990). The Council of Competitiveness found that Japan's standard of living showed 77 percent real growth between 1973 and 1993, compared to only 29 percent real growth for the American standard of living[2] (*Challenges,*

Figure 5.1
Percentage of Real Growth in the Standard of Living, 1973–1993

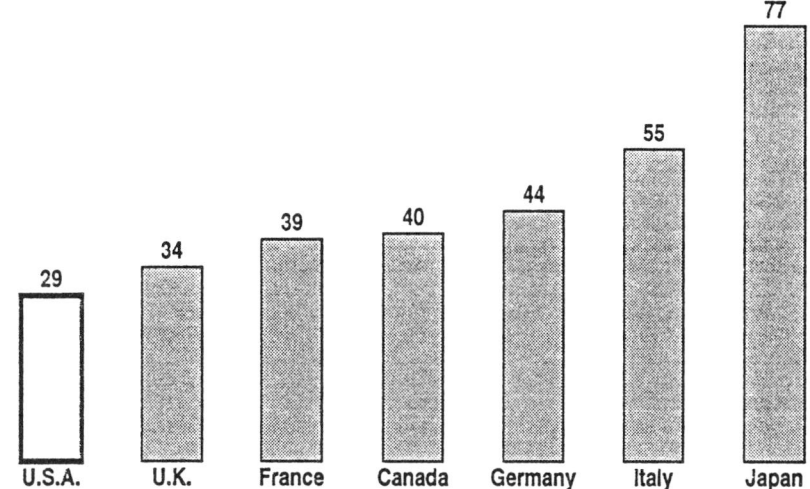

Note: "Standard of Living" defined as Gross Domestic Product per person.
Source: Council of Competitiveness, Challenges, July 1994.

July 1994). Many countries in Asia look at Japan's demonstrated success, consider the alien nature of many U.S. notions of economy, society, and development, and conclude that the way to riches is to close one's domestic market and export to America.[3] Washington policy makers are wrong to assume that the United States can decline to compete with Japan in markets and negotiations, and still win the minds and purchase orders of non-Japanese Asia.

Standards-setting in Japan, whether by government, industry group, or major company, is different from that in the United States. The chief differences include the fact that no Japanese standards-setting body, government, industry, or company is required to let anyone know that their business interests are being discussed, and you do not have effective legal recourse if you are injured.

A major cultural difference is that senior executives of Japanese multinationals are active on standards-setting bodies, whereas in the United States standards work is left to relatively junior management. NEC and Fujitsu, among others, have board members who have global standards coordination as a major part of their portfolio. Now, why do these senior executives bother?

To be sure, individuals active in standardization are honored by Japanese society. Standards panel members are appointed with an impressive document, bearing the vermilion official seal of a minister. Leaders of standards groups in

Japan are given medals by the MITI minister upon their retirement, or honored by the minister of Posts and Telecommunications at three thousand person receptions in Tokyo. Mr. Isao Yamashita, the first Japanese to become chairman of the ISO in 1986, was decorated by the Emperor for that achievement. But why does Japan care about standards? Mention the words "Japanese industrial policy" in the United States and the reflexive comment will be "picking winners and losers." This may be what French industrial policy is about, but Japanese industrial policy is different. Japan, unlike France, thinks of itself as a small, resource-poor nation alone in a world of jealous rivals, that cannot afford one false economic step.

Japanese industrial policy is about reducing economic risk to Japan. Technical standards reduce business risk. To mold a standard is to mold the future.

Japan has become infamous for trying to reduce business risk to its companies through the use of standards that attempt to replace foreign technology or burden foreign competitors. When we opened our office in 1984, American PBX makers complained that secret standards for voice quality, based upon an mechanical ear that existed only in Tokyo and Geneva, would prevent the selling of their superior equipment. More recently, American computer peripheral companies were howling about a proposed Japanese standard for uninterruptable power supplies that would act as a drag on American gains in market share.

But there have been some changes in the last ten years. More than twenty foreign-capital companies now participate in standards bodies affecting information technology (Figure 5.2). JIS drafting committees are open to foreigners and JIS Divisional Councils allow foreigners to offer an opinion. The Ministry of Posts and Telecommunications' top advisory committee that supervises standards, the Telecommunications Technology Study Council, seats representatives of American and EU-capital companies.

Indeed, there is increasingly conflict among Japanese companies themselves concerning unique Japanese standards. For example, in one debate, NEC and a Japanese trade association were strongly opposed to a suggestion that Japan establish required "intercompatibility" testing for equipment using ISDN technology. A senior NEC executive protested that it felt that producing the same equipment to EU specifications, to U.S. specifications, and to Japanese specifications unnecessarily raised its costs of development, and that Japan should allow equipment developed for foreign markets to be used in Japan. NEC was joined by the president of the Communications Industry Association of Japan, who pointed out that the telecommunications MOSS Talks between Japan and the United States had led to an agreement requiring only "no harm to the network," not "intercompatibility," as a condition for connecting to the public network. The CIAJ feared renewed trade friction.

Figure 5.2
JIS Standardization Procedures and Policies to Guarantee Their Transparency

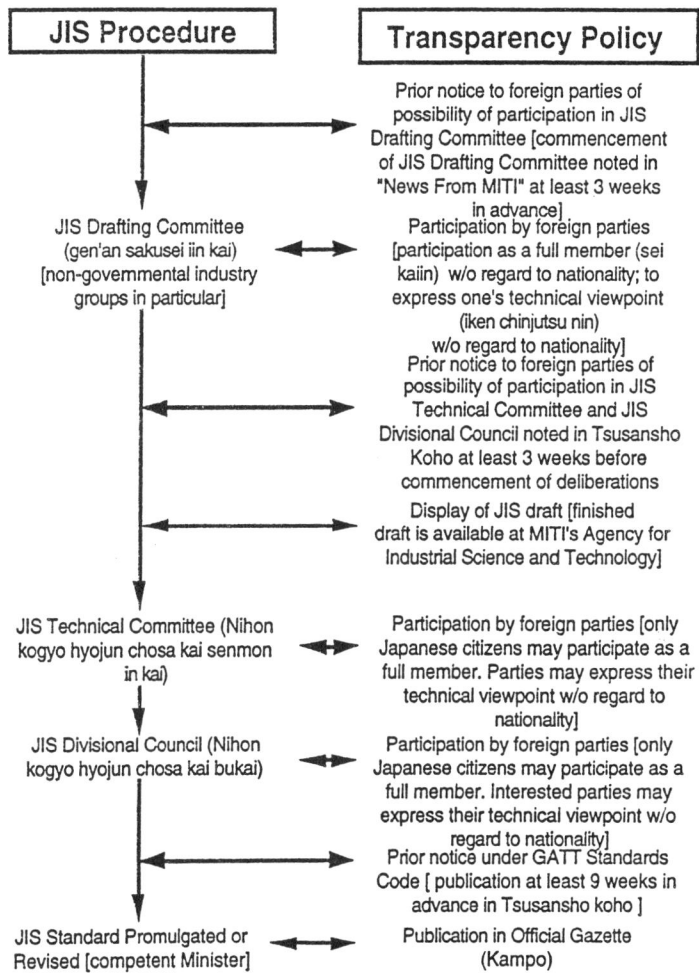

Source: Translated from a MITI document.

Fujitsu, on the contrary, championed unique Japanese standards. It was joined by a Japanese testing organization, which probably expected the monopoly for testing, and the Japanese government, which pointed out that the EU was subjecting Japanese companies to "intercompatibility" standards.

Some international standards are widely used in Japan; for example, the ISO 9000 series of quality-related standards, translated verbatim in October 1991, has been widely adopted by Japanese industry (Table 5.3). While the initial

Table 5.3
ISO 9000 in Japan: A 1992 Survey

Why Does Your Company Use ISO 9000?

In Order to Improve Internal Quality:	
Electronics	25.4%
Chemicals	13.4%
Other Manufacturing	9.8%
Steel	7.3%
Metal Fabrication	7.3%
For Domestic Business:	
Electronics	22.7%
Chemicals	16.7%
Other Manufacturing	10.7%
Precision Machinery	8.7%
Metal Fabrication	8.0%
For Overseas Business:	
Electronics	30.4%
Chemicals	18.4%
Machinery	8.0%
Precision Machinery	7.5%
Other Manufacturing	6.8%
For 312 Electronics Respondents:	
Number Using ISO 9000 Internally:	154%
Number Using ISO 9000 in Japan:	49%
Number Using ISO 9000 Overseas:	190%

Source: Hyojunka Journal, 1992 survey of 1377/2503 persons at JIS Seminar, January 17–31, 1992. (Response rate 55%)

impetus to ISO 9000 certification had been the fact that many EU customers required it as a condition for qualifying as a supplier, perhaps surprisingly in the land of the Deming Award, the ISO 9000 series has been embraced as a model for intracompany use and domestic transactions in Japan. In fact, hardly a week goes by without a Japanese electronics company taking out an advertisement, in the Japanese press, crowing about ISO 9000 certification of its factories.

ISO 9000 certification is big business. NEC, which has already achieved ISO 9000 certification for more than eighty of its facilities, has formed a new company, NEC Factory Engineering Inc., that reportedly plans to do ¥2.5 billion of business a year in ISO 9000 consulting (Nikkei Venture, September

1993, p. 80). Precision tool maker Mitsutoyo, Inc. reportedly expects to do ¥5 billion a year in ISO 9000 consulting and related tool sales (Nikkei Venture, September 1993, p. 80). Regardless of their merits as a quality system, the ISO 9000 series of standards will remain a factor in competing with Japanese companies in Japan and the EU for at least the next five years.

We have seen how Japanese standards can be a shield for Japanese companies or a sword for Japanese exports. They can also be a lifeline for American companies. For example, U.S. electronics companies currently have a majority share worldwide in the engineering workstation market. However, few U.S. workstation companies manufacture their own high-resolution displays or optical storage devices. The center of this technology is increasingly found in Japan. Even if American computer company X is justified in its confidence that its architecture is the best, or that its software is the best, it must ensure compatibility with the next generation of display and storage technology being developed in Japan. Unless Company X develops its new products with an eye to Japanese technical trends, it runs the risk of creating a new product that cannot take advantage of Japanese peripheral technology to the same extent as that of a competitor's product. To avoid this, several U.S. computer companies that seem to have no intention of entering the optical data storage industry as a manufacturer, nevertheless participate in Japanese optical data storage standardization procedures (Table 5.4). It is important to note that the major Japanese companies surveyed conduct a review of international technical standards before releasing the product on the market.

Industry standardization can suddenly unlock a potential market, leaving unprepared American companies in the path of massive Japanese competition. The American facsimile machine market fifteen years ago is a case in point. American-owned facsimile machine makers did not agree on a communications standard. Consumers were deterred by an expensive machine that could not communicate with anyone except the user of the identical make of facsimile. Japanese companies in 1981, however, adopted the CCITT GIII standard for facsimile machines, making it possible for one GIII machine to communicate with many different makes and models of other GIII machines. With some of the uncertainty over consumer acceptance of the facsimile reduced, Japanese fax makers began to mass produce and lower unit cost, leading to lower prices. Today, most offices in America use their fax machines constantly and the convenience of being able to send a fax without worrying about incompatibility is widely appreciated. However, there are not, to my knowledge, any American-owned manufacturers of commercial facsimile machines left in the United States. Sometimes, Japanese standards offer U.S. companies a brighter future. There are areas of information technology, such as CAD/CAM software, where U.S. companies have been successful in convincing Japan to base its domestic standard on American standards. In this

Table 5.4
Commercialization and Standardization in Japan

Japanese Company	Time from R&D to Market, Months	Domestic Technical Standards Considered	International Technical Standards Considered
Computers #1	12–24	Before R&D	Before R&D
Computers #2	12–24	During R&D	During R&D
Computers #3	12124	During R&D	During R&D
Computers #4	12–24	Before R&D	Before R&D
Computers #5	24–48	During R&D	During R&D
Telecoms #1	24–36	During R&D	During R&D
Telecoms #2	12–24	Before R&D	Before R&D
Telecoms #3	6–24	During R&D	During R&D
Telecoms #4	12–24	During R&D	During R&D
Consumer Elec. #1	6–12	Before R&D	Before R&D
Consumer Elec. #2	12–24	During R&D	During R&D
Semi. Mfg. Eq. #1	12–24	Before R&D	Before R&D
Semi. Mfg. Eq. #2	6–12	During R&D	During R&D

Source: British Chamber of Commerce in Japan, "Seihin-Ka," September 1988.

event, U.S. information companies obtain a lead over their Japanese competitors in making a sale.

For American information technology companies in Japan, it is still a major challenge to prepare one's corporate family to handle the Japanese standards-setting access opportunities that exist. The Japanese language is a hurdle: less than 1 percent of Japanese standards information is available in English. MITI makes summary information available in English regarding the more mature proposed standards, but other ministries publish only in Japanese, and all actual drafting and approval is in Japanese. There are few people in American corporate families whose job it is to understand both Japanese and English, the Japanese standards process, and to serve as liaison between the Japanese standards process and the American technical standards process. Sending the material to an outside translator, if a competent one can be found, will typically cost ¥8000 per page. A year's worth of standards can easily come to several thousand pages.

Lacking home office commitment, and therefore lacking a budget and staff, a number of American information technology companies in Japan have tried to use their sales engineers to participate in Japanese standards bodies. The results have usually been frustrating for both the participant and the company. A good salesman wants to close a deal and move on. A good standards

Figure 5.3
Mastering Japanese Technical Standards

- Employees participating in standard activities must believe that the company values their contribution as much as activities that contribute to the quarterly "bottom line."
- "Operating Division/Corporate" and "Staff/Line" branching of responsibilities and budget often starves standards activity and atomizes information going in/out of the company.
- Every employee should understand that the company values standards information affecting its business and should have access to a centralized database/electronic mail bulletin board on which to post information gathered.
- Hunters need decoys to attract the flock. Consider using foundations, think tanks, academics and journalists to obtain information that would not be given to you directly.
- Returning American academics, consultants, and career civil servants offer a wealth of information. Offer them an opportunity/forum for debriefing.
- Existing U.S. government technology-monitoring operations in Japan (State, NSF, Naval-Army-Air Force Research) should have their missions aligned to economic realities and should circulate their work to American industry.
- Support Congressional initiatives [Senate = Kerry (D-Mass.), House = Kolbe (R-Ariz.)] for a world-class U.S. trade center in Tokyo that would include standards experts.

participant is willing to follow a multiyear process and concentrate on the details. The sales engineer will often have his performance in the company rated by sales. While contacts made in the standards-setting process can be useful in marketing, sitting on a government committee is less likely to result in a sale than visiting a customer. Much more satisfactory results have been obtained when the American company places an R&D-oriented employee on a Japanese standards body.

Merely placing someone in Japan is not enough. Many people are needed just to follow standardization in Japan. That function must be supported not only in Japan, but also through liaison in the United States, Europe, and perhaps Asia as well. The timely transmission of business information worldwide is a skill that most Japanese manufacturers and many American service industries have mastered. I would not say that the same is true of American manufacturing in general.

James Morgan, the chairman of Applied Materials, stated that the Asian market grows by $3 billion per week, and now controls over 65 percent of the world's wealth.[4] For cultural and economic reasons, Japan would like Japanese telecommunications standards to become pan-Asian standards. Japan provides the majority of the budget of some Asian standards bodies, such as the Asia-Pacific Telecommunity. Japan provides the secretariat for others, such as the ASIA ISDN Council. Through 1993, Japan spent more than $500 billion in grants and $4 billion in government loans to build telecommunications infrastructure in Asia (*Denki tsushin*, March 1995, pp. 11–12).

The man who built early modern Japan, the shogun Tokugawa Ieyasu, followed a well-known proverb: "Go around if you are in a hurry." Japanese standards groups are careful to sound out their Asian neighbors' feelings about standardization. As a result, there is a willingness in Southeast Asia to be led by the Japanese, or at least by Asian, rather than American, standards.

How does an American company improve its ability to handle Japanese technical standards (Figure 5.3)? We have had eighteen thousand users of the American Electronics Association office since 1984, and worked with 487 U.S. electronics companies in Japan. This figure shows what I would suggest.

In Asia, confession is the first step toward rehabilitation. To the extent that a U.S. company feels that it has a problem with foreign market standards, it is on the road to remedying that problem. I look forward to the day when U.S. companies in Japan are serving as models not only of innovation, but also in guiding the future of the American electronics industry through standards participation.

NOTES

Official positions of the American Electronics Association are determined by its board of directors, which have not voted on the contents of this presentation.

1. The world semiconductor market is valued at $101.878 billion in sales.
2. Standard of living is defined as gross domestic product per person.
3. Indeed, China, the leader among "big emerging markets," announced in October 1994 that it would eliminate its trade deficit in electronics through an increase in exports of $150 billion by the year 2010 (UPI report of China Daily article, October 11, 1994).
4. Remarks to the Bay Area Council, February 11, 1993.

REFERENCES

Asahi Shimbun. June 14, 1990.
Council on Competitiveness, *Challenges*. July 1994.
Denki tsushin. March 1995, pp. 11–12.
Nikkei Venture. September 1993, p .80.
World Semiconductor Trade Service, 1994.

6

Technical Standards and Access to the Japanese Cellular Communications Equipment Market

Douglas J. Puffert

THE ECONOMICS OF COMPATIBILITY STANDARDS

National technical standards govern what products are accepted by the market in given countries. Standards may be divided into three, sometimes overlapping, categories: (1) standards of minimum acceptable product attributes in such dimensions as quality, safety, and environmental impact; (2) standards that define products in such terms as physical dimensions and product grades, facilitating business communication and consumer information; and (3) standards that assure a product's compatibility with other products used as parts of larger technical systems. Compatibility takes the form either of common physical interfaces ("plug-and-socket compatibility"), of common signal formats, or of information coding, in electronic or electromagnetic (radio) transmission or in information storage.

The principal issue for technical standards, in cellular communications networks, is the compatibility of electromagnetic signal formats between "network" equipment, primarily at the cell-site base stations, and "terminal" equipment, i.e., cellular handsets. Among the benefits of cellular standards, like other compatibility standards, are interoperability of equipment from different manufacturers, increased competition among manufacturers, and economies of scale in production.

A country's adoption of a particular standard may result in a competitive advantage for particular firms. Firms gain a competitive advantage when the standard embodies technologies in which those firms have either intellectual property rights (IPR), a research-based technical lead over competitors, or economies of scale in manufacturing as a result of supplying equipment for the same standard to other national markets.

National governments have sought to use technical standards to favor national firms in a number of different industries. For example, in early color television, wireline telecommunications equipment, and, more recently, high-definition television, governments have used national standards as hindrances to imports and have promoted adoption of their national standards in potential export markets.

Individual equipment suppliers pursue their own strategies with respect to standards. In many cases they form inter-firm alliances in an effort to establish particular technologies as "de facto" standards. A well-known example of this at the level of consumer choice is the VHS technology for video cassette recorders, which emerged as a standard over the rival Beta system. Firms try to do the same thing in cases where government bodies choose among competing systems. Often they have the support of their home governments in promoting their systems to foreign governments.

SYSTEM STANDARDS AND INTERNATIONAL COMPETITION IN CELLULAR COMMUNICATIONS EQUIPMENT: A GLOBAL PERSPECTIVE

The case of cellular communications equipment offers examples of a variety of corporate and governmental strategies. The two episodes of global competition in cellular systems correspond with the two generations of cellular technology: analog systems, introduced beginning in 1979, and digital systems, introduced beginning in 1992 (Table 6.1).

Analog Systems

Analog equipment suppliers and their national governments practiced three alternative strategies in global competition. The first strategy, possible at the time only in the United States and Canada, was to create a large, unified home market and use it as a springboard for exports. In 1982 the U.S. Federal Communications Commission (FCC) mandated nationwide adoption of a single system standard, known as Advanced Mobile Phone Service (AMPS). Subsequently, Canada adopted the same system. As a result of a large market with vigorous competition, U.S. and Canadian manufacturers improved their technology and realized substantial economies of scale, yielding high-quality and low-cost systems for export as well. Motorola used its expertise in AMPS to develop a similar system, Total Access Communication System (TACS), for export markets operating on a different part of the radio spectrum.

The second strategy, pursued by the Scandinavian countries and their major equipment suppliers, was to adopt a common standard, supported by multiple competing suppliers, in several countries. The result was the Nordic Mobile Telephone (NMT) system standard. As a result of fast-growing subscribership

Table 6.1
Global Cellular Communications System Standards

System Standard	Principal Countires or Regions	Date	Share of Global Subscribers
Analog Systems (share in early 1992)[a]			58%
AMPS	United States, Western Hemisphere, Asia & Pacific	1983	16%[b]
TACS	United Kingdom, Western Europe, Asia, Africa	1985	2%[c]
JTAC	Japan	1989	7%
NMT 450	Scandinavia, Western and Eastern Europe, Asia, Africa	1981	6%
NMT 900	Scandinavia, and	1986	6%
NTT	Japan	1979	6%
C-Netz	Germany	1985	4%
R-2000	France	1985	[c]
RTMS	Italy	1985	[c]
Comvik	Sweden	1981	[c]
Total			100%

[a]Share on the eve of introduction of digital service. NMT and C-Netz shares have fallen substantially in the intervening period.
[b]The share for JTAC is included with the share for TACS.
[c]Less than 0.5 percent.

Source: Compiled by the author.

in Scandinavia, vigorous promotion of the system in additional countries, and competition among suppliers, NMT equipment improved substantially in quality and price and remained competitive with AMPS and TACS in new markets.

The third strategy was to adopt country-specific standards with equipment supplied by "national champion" firms. This strategy, adopted by Germany, France, Italy, and Japan, was not very successful. With relatively small home user bases and a lack of competition in home markets, equipment quality and prices did not improve enough to compete with other systems in potential export markets.

Most analog network equipment suppliers have been competitive only in systems that they have had experience in producing for their home markets. The principal exceptions to this have been the Swedish firm Ericsson which was able to transfer its expertise from NMT to AMPS and TACS, and the U.S. firm Motorola, which invented TACS and has supplied numerous TACS systems. NEC (Japan) offers AMPS and TACS systems but has had little success in the markets for these products. In terminal equipment, by contrast, Japanese firms have had substantial success in selling products for a variety of system standards. It appears that manufacturing technology has been much more important than system-specific technology in developing competitive terminal equipment.

Digital Systems

Both Europe and Japan learned lessons from the analog equipment market, which they applied to the digital market. European governments and firms were committed to a common standard so that a large, unified and competitive market would result in quality and cost improvements that would make European suppliers competitive in both European and world markets. The name of the standard, originally Groupe Speciale Mobile (GSM), was changed to Global System for Mobile Communications (still GSM) in order to reflect an increasing emphasis on promoting the standard as a de facto world digital cellular standard. GSM has since been adopted by a large number of countries.

The Europeans also protected their standard with IPR and sought to use IPR licensing terms to strengthen the position of European firms against potential foreign competitors. In the event Motorola was able to claim IPR for its own because of technology embodied in GSM—European courts have disallowed discriminatory licensing provisions—so Motorola and other non-European firms appear to be in a position to compete aggressively in markets for GSM equipment.

In the United States, the FCC declined to mandate a digital cellular standard, but equipment manufacturers established one themselves, commonly known as Time-Division Multiple Access (TDMA), in 1989. However, after this date, one U.S. equipment firm developed an alternative system, known as Code-Division Multiple Access (CDMA). As a result, different U.S. cellular service providers have adopted different digital systems. United States equipment manufacturers appear likely to realize lower economies of scale than they otherwise might have, and U.S. TDMA equipment manufacturers are reportedly at some disadvantage in competition with GSM suppliers in export markets.

The Japanese, meanwhile, adopted a digital cellular system similar to the U.S. TDMA system, expecting that system to account for as much as 65 percent of the world market in digital cellular systems. That expectation now appears unrealistic.

STANDARDS AND MARKET ACCESS IN JAPANESE CELLULAR EQUIPMENT

Analog Equipment

Although Japan was the first country with cellular telephone service, its service prices (which included handset rental) were until recently relatively high and its rate of subscribership relatively low, partly because Japan used its own cellular standard. Motorola saw an opportunity for profit in Japan, particularly if it could sell TACS network and terminal equipment. As a result of lessons learned in trying to sell pagers in Japan, Motorola launched a two-pronged strategy to enter the market. First, Motorola enlisted the support of the U.S. Trade Representative (USTR) in removing trade barriers and secondly, Motorola formed an alliance with an entrepreneurial Japanese firm, DDI, which proposed to develop a second nationwide cellular network in Japan.

As a result of USTR involvement, the 1985 Market-Oriented Sector-Selective (MOSS) agreement on telecommunications between the United States and Japanese governments affirmed U.S. firms' access to the Japanese cellular equipment market and specified that a "North-American type" cellular system (i.e., AMPS or TACS) could be used in Japan. Meanwhile, DDI and Motorola met an unexpected obstacle to their plans for a nationwide TACS network when a rival service provider, IDO, was formed and Japan's Ministry of Posts and Telecommunications (MPT) awarded it the franchise for the populous eastern half of the country, leaving only the west for DDI. Subsequently DDI's region was enlarged, but the company was still excluded from the Tokyo-Nagoya region, the commercial heart of Japan. Because IDO adopted the standard of NTT, Japan's major telephone service company, there was no technical means for a DDI subscriber to use a TACS phone when traveling to the Tokyo-Nagoya region.

DDI and Motorola protested that this arrangement put DDI and TACS at a competitive disadvantage relative to NTT and its system. As a result of further USTR involvement, a 1989 governmental agreement specified that a North American type cellular network would be developed throughout Japan, and IDO was required to allocate 20 percent of its radio spectrum capacity to the TACS system. However, IDO was given little incentive to comply quickly with this requirement, and it proceeded very slowly in installing TACS equipment, which it ordered from Motorola.

As a result, in 1994 the USTR found Japan to be in violation of both the 1985 and 1989 agreements. This led quickly to a new agreement requiring IDO to actively promote TACS service and to offer TACS in 95 percent of its service area by September 1995. Reportedly, IDO is now deploying TACS ahead of the agreed schedule.

The 1994 agreement also provided that Japanese cellular service suppliers would henceforth permit customers to buy their own handsets rather than lease them from service providers. This has reportedly led to more vigorous competition in both handsets and service, bringing about substantial declines in prices and a large increase in subscribership for each of the networks.

Digital Equipment

Access to the market for Japanese digital cellular equipment has been relatively open from the beginning. As noted above, Japan designed its standard, JDC, to be similar to the U.S. TDMA system, exposing Japan's equipment suppliers to competition in Japan while also, it was hoped, preparing them to compete more effectively in foreign markets for cellular network equipment. Motorola and, to a lesser extent, other foreign firms were directly involved in designing the standard, and Motorola's voice coding technology was incorporated into the standard, as it had been into TDMA. Motorola, Ericsson, and AT&T were selected at an early date as suppliers of network equipment. Motorola has also become a major supplier of JDC handsets. In the year since digital service began, subscribership has grown rapidly.

7

Introducing a Standard for Digitalized Medical Images in Japan

Aki Yoshikawa

In Japan, the rapid spread of high technology medical equipment is striking. Japanese per capita diffusion of such technologies is much greater than that of any other industrialized nation, including the United States.[1] In particular, the widespread diffusion of medical imaging machines (i.e., CT [computed tomography]; MRI [magnetic resonance imaging]; and CR [computed radiography]), has increased demand for the storage and on-line exchange of medical images. In Japan, this demand led to the introduction of a standard method for storing medical images called IS&C (Image Save and Carry) that stores medical images in magneto-optical disks (MODs). This chapter analyzes the introduction of IS&C and the disputes surrounding its introduction.

MEDICAL IMAGING TECHNOLOGY

Established: CT and MRI

The diffusion of medical imaging machines such as CT and MRI has been dramatic in Japan. The spread of CT (Figure 7.1) has been considerably higher in Japan than in any other industrialized nation (Yoshikawa et al., 1991). Per capita, Japan has more than twice the number of CTs than the second highest nation, the United States. In the United Kingdom, where CT was invented, there are fewer CTs per capita than in any other industrialized nation. The rapid spread of CT scanners in Japan was partially fueled by competition among hospitals.[2] Japan has over ten thousand hospitals and 1.6 million beds, while the United States, with double Japan's population, has fewer than six thousand and five hundred hospitals and 1.2 million beds.

Figure 7.1
International Comparison of CT Diffusion

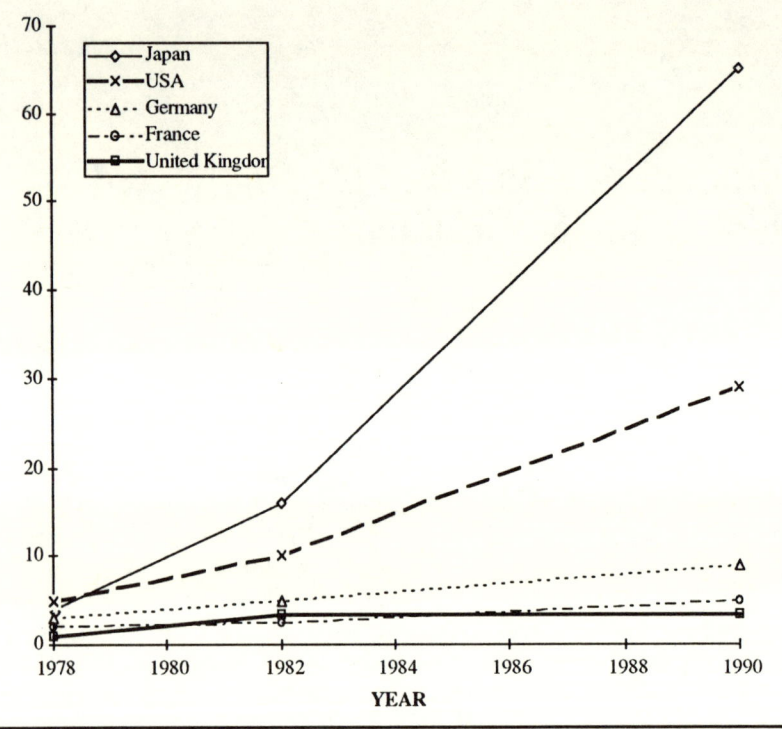

Source: Author's own estimates, based on Japan-Koseisho Surveys.

According to a survey in 1993, there are 10,385 CT (Figure 7.2) and 1,683 MRI (Figure 7.3) in Japan (*Yoshikawa*, 1994). Toshiba Medical maintains the largest share in both markets: 44.0 percent in CT and 26.5 percent in MRI (*Iryokiki Hakusho*, 1994). Hitachi has the second largest share with 22.9 percent in CT and 19.7 percent in MRI. Yokogawa Medical (GE) has the third largest share, 20.4 percent in CT and 14.2 percent in MRI (*Iryokiki Hakusho*, 1994).

Newcomer: CR

CR (computer radiography) is a relatively new technology, but one that is spreading rapidly as the medical world realizes its advantages and potential. Instead of using traditional X-ray film, CR produces digitalized medical images using imaging plates, thus allowing doctors to analyze the images with

Figure 7.2
CT Market in Japan

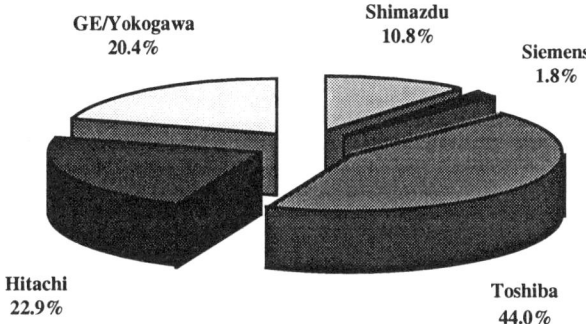

Source: Iryokiki Hakusho 1994 (Tokyo: Shin Iryo, 1994).

Figure 7.3
MRI Market in Japan

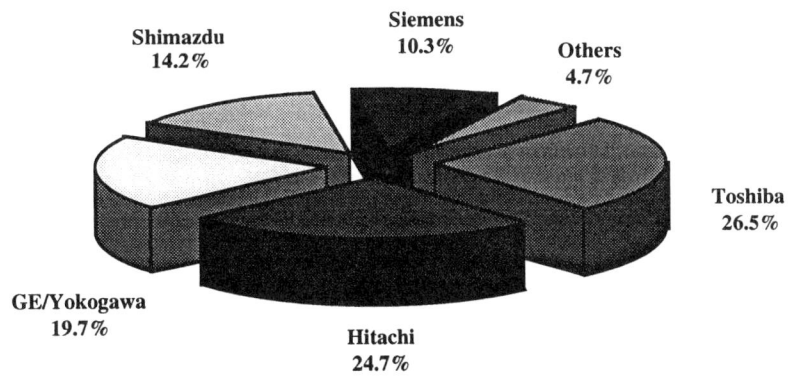

Source: Iryokiki Hakusho 1994 (Tokyo: Shin Iryo, 1994).

Figure 7.4
Diffusion of CR in Japan

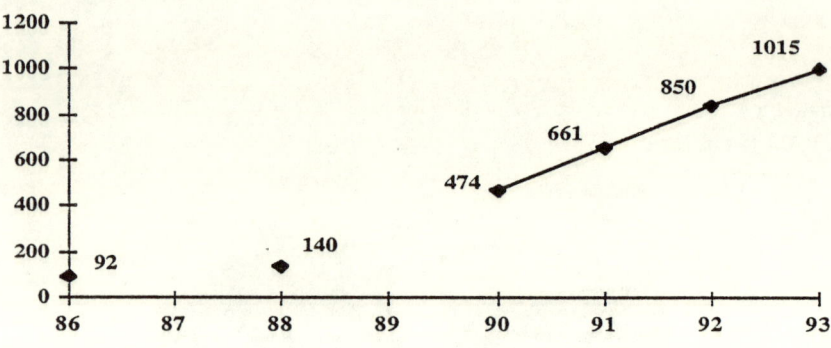

Source: Iryokiki Hakusho 1994 (Tokyo: Shin Iryo, 1994).

computers. In this sense, CR is a superior substitute for traditional X-ray machines.

CR was invented by Fuji Photo and its market is essentially monopolized by its subsidiary, Fuji Medical. So far, the company has effectively blocked the entry of other companies into the CCCR market. The sale of CCCR (Figure 7.4) began in 1983; by 1993, there were 1,015 CR in Japan (*Yoshikawa*, 1994).

The invention of CR is a mixed blessing for Fuji Photo, which also maintains the largest share in the Japanese X-ray film market. While many believe that the market for CT is saturated, and the market for MRI is nearing saturation, the market for CR is expected to expand as CR replaces traditional X-ray machines. CR is still a relatively expensive technology with a low-end model (FCR AC-3) advertised for ¥13,000,000 (*Shin Iryo*, October 1993, p. 149). A high-end model (FCR9501) costs ¥37,000,000 (*Shin Iryo*, March 1994, p. 109). Including the peripherals, a top-of-the-line CR system is priced at ¥60,000,000 (*Shin Iryo*, May 1993, p. 134). Meanwhile, Japan's X-ray film market is characterized as a mature market, estimated at ¥110 billion[3] (*Shin Iryo*, July 1993, p. 112) with an annual growth of 2–3 percent[4] (*Shin Iryo*, January 1993, p. 70). The market is shared by several major domestic (Fuji, Konica) and foreign producers (Kodak, Dupont, 3M). Because Fuji is essentially monopolizing the CR market and selling approximately two-hundred CR machines every year, CR is a strategically important product for Fuji.

A Standard for Digitalized Medical Images

Figure 7.5
PACS in Japan

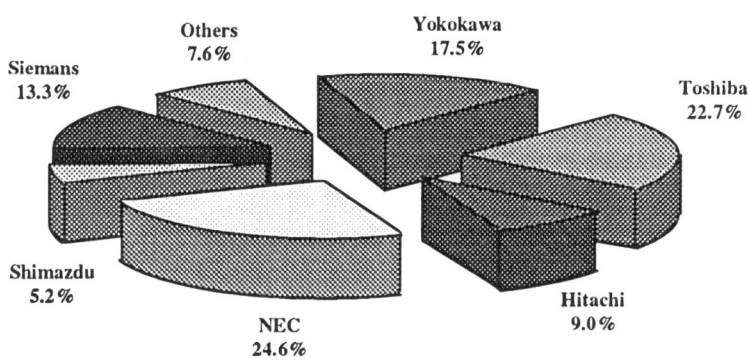

Source: *Iryokiki Hakusho 1994* (Tokyo: Shin Iryo, 1994).

STORING AND COMMUNICATING MEDICAL IMAGES WITH PACS

Japanese hospital information systems (HIS) were developed when medical facilities found them useful in preparing health insurance claims.[5] In 1980, only 6.5 percent of Japanese medical facilities used computers to prepare health insurance claims, accounting for 11.8 percent of the total number of claims filed. As evidence of HIS's popularity,[6] by 1990, 43.6 percent of medical facilities were filing claims on computers, increasing the percentage of claims filed by computers to 68.7 percent (Yoshikawa). Looking at hospitals[7] alone, 81.6 percent used computers to prepare insurance claims in 1990, accounting for 93.3 percent of all claims filed.[8] Recently, in an effort to expand the use of HIS to other applications, hospitals have begun integrating the computer system used for accounting with laboratory information and pharmaceutical inventory.

With the great popularity of medical imaging in Japan, an information system to store and communicate digitalized medical images, known as Picture Archiving and Communication Systems (PACS), was introduced in the early 1980s and was expected to spread quickly. The number of such systems in Japan (Figure 7.5) has increased from 23 in 1987, to 96 in 1990, to 211 in 1993 (Yoshikawa). However, unlike the large scale, fully networked and integrated PACS in the United States most Japanese PACS are "stand-alone," without network capabilities and integration with other hospital systems (Yoshikawa).

Integrating PACS with existing HIS is a costly process for hospitals, requiring the purchase of hundreds of high resolution terminals to display and transmit visual information. Because of the expense, only the government can finance such integrated PACS. Given the traditional government subsidies to university hospitals, it is not surprising that the first fully integrated PACS were installed at Hokkaido University in June 1989. The Hokkaido system was jointly developed with NEC. Various prototype PACS have also been installed in other university hospitals.

The prevalence of "stand-alone" PACS in Japan may be a result of the marketing efforts by manufacturers of medical imaging equipment and the equipment's popularity. Many large hospitals purchased these "stand-alone" PACS with their medical imaging equipment. Toshiba and Hitachi are two Japanese manufacturers who offer such complete packages, including medical imaging products and laboratory automation systems. These companies have a clear competitive advantage over those who cannot provide such packages.

Although it was widely speculated that PACS would be used in many hospitals, the diffusion of these systems is far less spectacular than had been anticipated. There are several reasons for this (Minato, 1994). For doctors and hospitals, the advantage and convenience of using digitalized storage and communication systems has yet to justify their purchase. In addition, manufacturers failed to produce the desired products. Finally, the legal framework for maintaining patients' medical images in digital format was not well established until the spring of 1994.

INTERNATIONAL STANDARDS FOR DIGITAL COMMUNICATION OF MEDICAL IMAGES

In 1982, the American College of Radiology (ACR) and the National Electrical Manufacturers' Association (NEMA) formed the ACR/NEMA Committee to establish digital communication standards for medical imaging. Their first version (ACR/NEMA v.1.0) was released in 1985, followed by the second version (ACR/NEMA v.2.0) in 1988.

In 1989, the ACR/NEMA v.2.0 standard was adopted by the Japan Industries Association of Radiation Apparatus (JIRA). In Japan, it was know as the MIPS (Medical Imaging and Processing Systems) Standard 87 and 89. It was natural for Japan to adopt the ACR/NEMA standard because many Japanese researchers were familiar with it through their training and experience in the United States. The only significant difference between the ACR/NEMA and MIPS standards is the inclusion of Japanese characters in the MIPS standards. The fact that both Japanese and American manufactures have large shares in both countries made it pragmatic for manufacturers to support a common standard.

In 1994, the ACR/NEMA v.2.0 standard was revised extensively and renamed DICOM (Digital Imaging and Communications in Medicine) v.3.0. DICOM is a voluntary standard concerned with the exchange of medical images. The DICOM Committee, whose purpose is to develop and promote the use of the DICOM standard, is located within NEMA. The DICOM committee includes various academic societies: the ACR, the American College of Cardiology, the American Urological Association, the American Society of Gastrointestinal Association, the American Academy of Ophthalmology, and other private companies. The wide variety of members in the DICOM committee indicates that future DICOM applications may extend to medical fields outside of radiology (Kajiwara, 1995).

DICOM became a European standard in 1994 when the Comité Européen de Normalisation (CEN) issued a prestandard based on DICOM. The Cancer Research Center in Japan, with the cooperation of IBM, NEC, and Andersen Consulting, developed a protocol standard based on DICOM. Today, DICOM is the de facto international standard for the network exchange of medical images. This standard allows diagnostic images to be stored and transmitted via computer, eliminating the need for storing and transmitting images on film.

The Japanese Standard: IS&C

Frustrated by the slow development of integrated PACS in Japan, some observers believed that the introduction of an off-line system of communication would be easier and strategically more important than the introduction of an on-line system. Based on this view, Japan adopted a more pragmatic and less ambitious approach to communicating medical images. They developed an off-line system to share digitalized medical imaging information. Image Save and Carry (IS&C) is the Japanese off-line standard that reads images from a removable computer disk.

IS&C (pronounced *Isaac*) was introduced in 1989 by Professor Oyama of the Tokyo Institute of Technology. The product stores images from CT, MRI, and CR machines onto a 130 mm MOD (magneto-optical disk). The MOD can then be transported between medical departments and hospitals when the images are requested. IS&C is off-line rather than on-line based technology. The images stored in MODs can be accessed and searched at workstations. MODs are similar to portable patient records—following a patient to various medical facilities (Figure 7.6).

IS&C was developed by the IS&C Committee, comprised of members of the Japan PACS Committee (JPACS) and the Medical Information System Development Center (MEDIS). JPACS is an industrial organization created by companies involved in the medical X-ray field. MEDIS is a semi-public corporation (Zaidan Hojin) created in 1974 by Japan's Ministry of Health and Welfare (MHW) and the Ministry of International Trade and Industry (MITI). Private companies involved in the IS&C Committee include both domestic (Canon, Epson, Fuji Photo, Fujitsu, Hitachi, Konica, NEC, NTT Data,

Figure 7.6
IS&C Concept

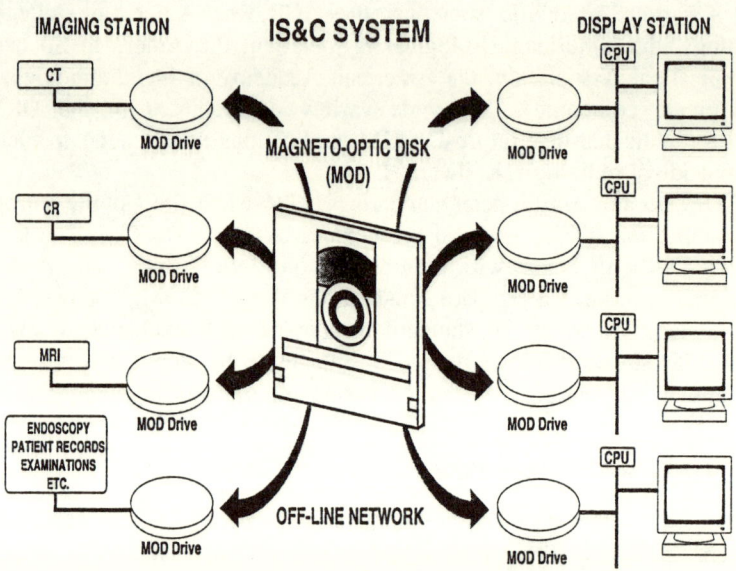

Olympus Optical, Ricoh, Seiko, Shimadzu, Toshiba, etc.) and foreign subsidiaries (IBM, Kodak, Siemens, and Yokogawa/GE).

The MHW had previously required that all radiographic images be archived on photographic film. However, the MHW recently changed the regulations to allow images to be archived on a digital medium instead of film, if the digital system met specific requirements. An official notification issued in March 1994 by the director general of the MHW's Health Policy Bureau (Kenko Seisaku Kyoku) opened the door to electronic storage of medical images, subject to the three requirements: (1) reproducibility of data, (2) security of data, and (3) interoperability of data (data can be read on all machines) (*Koseisho Kenko Seisaku-Kyokucho Tsuuchi*, March 29, 1994).

With this notification, the MHW indirectly encouraged the industry to develop standards for storing information in digital format. This format may one day replace film as the archive medium. MEDIS has already established the first electronic image archiving standard, following the above three criteria. This standard, know as Standard I *(Kyotsu Kikaku I)*, is based on the IS&C standard.

The IS&C standard is used exclusively for record keeping and has a mechanism to ensure data security. This is unlike the DICOM standard, which does not have a security mechanism yet. This problem had been identified by DICOM and IS&C researchers. Thus, the Japanese effort in creating a security algorithm in the IS&C system is therefore rational and well justified.

However, even with the official endorsement of digital technology as a storage medium for medical images, the issues of cost, convenience, and quality still remain and must be solved before the use of digital technology can become widespread. First, storing X-rays on film is far less expensive than storing images in digital format. Currently, one MOD costs about ¥30,000 and stores eighty X-rays (*Iryo to Computer* [Special IS&C edition]). Therefore, each X-ray image costs ¥375 to store on an MOD. Compared to traditional film, this is a more expensive option. Price needs to be competitive before MOD can compete with film. The government, through modification of the fee schedule, can create the financial incentive for doctors and hospitals to use digital technology.

Second, doctors may still prefer traditional film because of its familiarity and convenience. Although MODs save a large amount of space (one thousand, two hundred CT images can be stored on one MOD), doctors may find them inconvenient. Doctors can only access the images at a workstation, and even then, the process of displaying each image on the computer is painfully time consuming, taking more than ninety seconds. For specialists who look at hundreds of images a day, this is an impractical option. Another drawback to computer images is that their resolution is much poorer than X-rays. Often, doctors rely on sharp resolution to diagnose patients, thus making computer images ineffective. In sum, the convenience and quality of digital technology is not yet sufficient for the needs of many doctors.

IMPLEMENTATION AND STANDARD I (IS&C)

The government introduces a standard hoping that with positive network externalities, a product will become more convenient to use and thus more valuable to users as more and more users adopt compatible products. The widespread use of compatible products helps suppliers realize economy of scale in production, creating products that can be sold at lower prices and perhaps with higher quality. At a reduced price, the demand for products increases. MEDIS attempted to achieve this positive chain of events by introducing a standard for medical imaging.

MEDIS originally planned to introduce only one standard, the IS&C based Standard I. The original scheme envisioned by the Japanese government and MEDIS was extremely restrictive. Under the scheme, MEDIS would be responsible for maintaining the quality of the products and ensuring that they met MHW criteria. MEDIS would contract companies interested in developing products based on the IS&C standard. Part of the contract would give a company access to security codes and allow it to purchase medical disk drives from drive manufacturers. The company would agree not to disclose the security information to anyone and to sell the products only to licensed Japanese medical practitioners.[9] MEDIS would then certify each product satisfying the MHW criteria. Although the contract would be available to companies around the world and would allow products to be developed in any country, the products would have to be sold in Japan.

Foreign companies feared this "semi-official" certification by MEDIS. While the government maintains that this certification would not be a requirement in government procurement of foreign products, it would become an implicit requirement of both government and nongovernment procurement if the certification system is initiated. Regardless of what the official policy is, users would naturally turn to products that had this certification. Thus, the certification system is a "nongovernment level of regulation." Some industry insiders criticized that by designating MEDIS as the regulator, the MHW puts an industry organization between itself and foreign companies.

Introducing Standard II (DICOM)

MEDIS had maintained that Standard II, based on DICOM, would be introduced sometime after the introduction of Standard I. Security measures would have to be added to the DICOM standard before Standard II could meet MHW criteria for image archiving. The DICOM Committee and MEDIS held joint meetings to facilitate such work. Even though no significant technical difficulties were expected, the development of Standard II lagged behind Standard I, which had already been studied by MEDIS for some time. On Nov. 8, 1994, MEDIS published the IS&C based Standard I. At the time, certification of an alternative system (DICOM based Standard II) was not expected soon.

When MEDIS's announced its intent to introduce an alternative standard based on DICOM, the news was met with some dismay because it created an environment of two incompatible systems in Japan. The decision to introduce the DICOM based alternative was primarily due to criticism voiced by researchers. They pointed out that DICOM is the de facto international standard, not IS&C.

CRITICISM OF IS&C

IS&C is facing tough criticism both in Japan and abroad. As mentioned, Japanese researchers who are familiar with the DICOM system widely acknowledge DICOM as the de facto international standard. Thus, the most common criticism focuses on the fact that the new off-line standard should be DICOM based, instead of creating an unique Japanese standard like IS&C. On-line communication of medical images is the trend of the future. Any off-line standard should be easily compatible with the de facto international on-line standard, DICOM.

Dr. Kotaro Minato of Kyoto University Medical School states that 130 mm magneto-optical disks are outdated (Minato, 1994). The IS&C standard recommends only one medium for data storage, the 130 mm MOD. Since 1989, several other storage mediums have been developed (e.g., digital video disk [DVD], optical discs, and magnetic tape). However, these are not included in the IS&C standard. Thus, the proposed IS&C standards do not appear to have a mechanism for including new technology in a rapid and equitable manner. The standards appear to have been developed based solely on 1989 technology. IBM, once the leader[10] in promoting MOD, recently announced its withdrawal from MOD drive production by the end of August 1995 (*Nihon Keizai Shinbun*, August, 10 1995). Toshiba, targeting the development of DVD, has already withdrawn from MOD drive production, while Matsushita and Sony are targeting optical disc development (*Nihon Keizai Shinbun*, August 10, 1995).

In the information systems field, characterized by daily changes, relying on 1989 technology is likely to be a critical mistake. Dr. Ken'ichiro Kajiwara of Kurume University Medical School expresses his opinion against IS&C, claiming that its reliance on a single medium (MOD) is passé (Kajiwara, 1995). Professor Kajiwara also criticizes the development of IS&C, claiming that too few clinical viewpoints were considered during the process. Based on my interviews, IS&C is facing opposition from those who are familiar with the system. Many Japanese are in favor of the international DICOM-based standard for data storage rather than their own IS&C.

Foreign companies also object to the IS&C based Standard I in light of the nontariff trade barrier that it would create. Foreign companies argue that establishing a Japanese specific standard (IS&C) and making foreign entry difficult will separate the Japanese market from international competition. The Japanese market would be a safe haven for domestic producers. All non-Japanese companies currently utilize the DICOM standard in developing storage devices for digitalized medical images. Adoption of the proposed policy by the Japanese government would effectively end foreign sales of many types of diagnostic imaging equipment in the Japanese market. The IS&C based Standard I would make it impossible for U.S. companies with alternative technology to successfully compete in the Japanese market. Innovative technology would have no commercial opportunities in the Japanese market with this exclusive technical standard. Hence, foreign companies demanded the delay of the introduction of Standard I and a simultaneous introduction of Standard II (DICOM).

SIMULTANEOUS INTRODUCTION OF STANDARDS I AND II

The Compromise

Although MEDIS had released Standard I in Fall 1994 and had planned to make it official in Spring 1995, due to foreign criticisms (*gaiatsu*), MEDIS postponed its introduction and certification of Standard I. Early in the spring of 1995, reports claimed that MEDIS was adamant about introducing the IS&C based Standard I prior to the DICOM based Standard II. So, in late June 1995, MEDIS surprised many when it announced that it would abandon its original plan and would introduce Standard I and II together. Standard II would be a DICOM based standard with an additional security algorithm, alleviating the earlier concerns over DICOM security.

Foreign companies viewed this as a victory against the Japanese government. However, since the storage medium for both the IS&C based Standard I and the DICOM based Standard II are restricted to 130 mm MOD, it is only a partial victory for foreigners.

New Problems

The simultaneous introduction of Standard I and II creates its own problems. The situation is similar to that of the simultaneous existence of Beta and VHS video recording formats in the past. The coexistence of two different, incompatible standards will undermine the fundamental reason for having a standard. Furthermore, one of the three requirements set by the MHW, the interoperability of the data, also cannot be obtained.

The coexistence of two standards will also make the harmonization of the two difficult in the future. Eventually, through competition in the market, there will be a winning standard and a losing one. The process of identifying the winner and loser will take longer with two official standards. Unless Japanese producers have a very strong incentive to continue developing and producing IS&C based products, it is reasonable to expect that Standard I will lose. With the simultaneous introduction of both IS&C and DICOM based standards, Japanese producers do not have a safe haven for IS&C based products. Instead, both Japanese and foreign producers will likely concentrate on developing DICOM based products that can be marketed globally.

Thus, IS&C faces an uphill battle. There is still the possibility that IS&C will dominate the Japanese market. If Japanese producers are successful at turning out many IS&C products, Japanese hospitals, with little need for on-line communication, may purchase them and perhaps create a separate market from the rest of the world. It is probable that Japanese companies have progressed far enough in developing IS&C products and would have a competitive advantage with an early introduction.

A Standard for Digitalized Medical Images 105

Unfortunately, even if this remote possibility of IS&C success becomes a reality, this success would only be for the short term. Such dominance would likely be terminated, again, by gaiatsu. Although the majority of Japanese hospitals are in the private sector, among large hospitals (those with more than three hundred beds), the public sector dominates. Large public hospitals are the ones which are most likely to digitalize their medical information storage first. With the prominence of public sector medicine, and therefore large government procurement, the medical equipment market is a likely target for gaiatsu pressure.

IN RETROSPECT

The IS&C experience illustrates the difficulty of introducing a rigid technical standard in an area of rapid technological change. Imposing a specific standard may hinder growth of the industry and reduce diversity if inappropriate or outdated technology is selected as the standard.

Although there was a pragmatic and realistic effort to develop an off-line standard, MEDIS underestimated the technological advancement and the growing demand for on-line based technology. It seriously underestimated the dominance of DICOM in the global market when it chose IS&C for its standard. This leads to the important issue of how standards should be chosen and promoted. Looking at IS&C and DICOM gives a good contrast. With IS&C, the approach was top down. MEDIS, a semi-public entity organized by the government, chose the standard based mostly on government and industrial association advice. With ACR/NEMA and DICOM, standards were set jointly by users (various academic societies) as well as industrial associations. Japanese academic medical societies are relatively young and sometimes their input is not sufficiently included as the IS&C case exemplifies. In comparison with ACR/NEMA, DICOM, and MIPS, the IS&C decision seemed much more supply-side oriented.

With IS&C, the Japanese introduced security to the communication of medical images, which is a noble undertaking. DICOM was also looking for a way to establish a security mechanism that would have greatly contributed to the development of the de facto international standard. Fundamentally, IS&C and DICOM were striving for the same improvement to the system. Thus, the cooperation of IS&C and DICOM could have been promoted by the establishment of a forum in which IS&C and DICOM specialists would have the opportunity to discuss common problems and seek mutual solutions. A lack of communication seems, in part, to blame for the foreign distrust of IS&C.

The IS&C experience also shows the relative naiveté of MEDIS, the semi-public foundation headed by retired bureaucrats, toward the views of foreign companies and gaiatsu. The decision to introduce the two different standards simultaneously shows that the Japanese government is sometimes susceptive toward foreign pressure.

NOTES

1. For more information about the Japanese Health care system, see Okimoto and Yoshikawa, eds., Japan's Health System: Efficiency and Effectiveness in Universal Care (Faulkner & Gray, 1993).
2. For an empirical study of diffusion of medical diagnostic technology in Japan, see Vogt et al., "The Role of Diagnostic Technology in Competition among Japanese Hospitals," International Journal of Technology Management (summer 1995).
3. Statement by Akihiko Nakane, director of Konica.
4. Statement by Isao Kozu, CEO of Fuji Medical System.
5. Laboratory automation systems (e.g., automatic examination of blood specimen) and pharmacy automation systems (e.g., drug dispensing and inventory control) are the next target for HIS.
6. Data are from Shakai Hoken Junpo (12/1/91), p. 5.
7. Hospitals are defined as medical facilities with twenty or more beds.
8. Data are from Shakai Hoken Junpo (12/1/91), p. 5.
9. The DICOM standard, if adopted as an official international standard, would be an open standard. In contrast, Japanese government officials have told European standards officials that they have no intention of licensing or otherwise making available IS&C technology to non-Japanese firms.
10. IBM has been the leading promoter of MOD, especially 3.5-inch drives. The main reason for IBM to withdraw from the MOD drive production is its high price compared to competing products, such as hard disk drives.

REFERENCES

Iryokiki Hakusho. Tokyo, Japan: Shin Iryo, 1994.
Kajiwara, Kenichiro. "DICOM to Telemedicine no kanousei." *Shin Iryo* (July 1995): 58–60.
Koseisho Kenko Seisaku-Kyokucho Tsunchi. "X-sen tou no Kojki Disk ton heno Hoson ni tsuite." March 29, 1994.
Minato, Kotaro. "PACS no Suishinyoso to Sogaiyoso." *Shin Iryo* (August 1994): 46–49.
Nihon Keizai Shinbun. August 10, 1995.
Shakai Hoken Junpo. December 1, 1991, p. 5.
Shin Iryo. March 1994, p. 109.
———. October 1993, p. 149.
———. July 1993, p. 112.
———. May 1993, p. 134.
———. January 1993, pp. 70.
Yoshikawa, Aki, Norihiko Shirouzu, and Matthew Holt. "How Does Japan Do It?: Doctors and Hospitals in a Universal Health Care System." *Stanford Law and Policy Review* (fall 1991): 111–137.

8

Critical Success Factors Leading to Japanese Dominance in Electronics Packaging

Michael J. Kelly

INTRODUCTION

This chapter is based upon a report entitled *Electronic Manufacturing and Packaging in Japan*, prepared for the Japanese Technology Evaluation Center. The methodology of JTEC is to organize a team of U.S. experts who study and visit Japanese industrial and research sites and then report their findings.

This report summarizes the status of electronic manufacturing and packaging technology in Japan, and its impact on competition in electronic manufacturing in general. In addition to electronic manufacturing technologies, the report covers technology and manufacturing infrastructure, electronics manufacturing and assembly, quality assurance and reliability in the Japanese electronics industry, and successful product realization strategies. It indicates how technological dominance goes hand in hand with standards setting dominance.

The panel found that Japan leads the United States in almost every electronic packaging technology. Japan clearly has achieved a strategic advantage in electronics production and process technologies. The JTEC panel members believe that Japanese competitors could be leading U.S. firms by as much as a decade in some electronics process technologies. Japan has established this marked competitive advantage in electronics as a consequence of developing low-cost, high-volume consumer products. Japan's infrastructure, and the remarkable cohesiveness of vision and purpose in government and industry, are key factors in the success of Japan's electronics industry. Although Japan will continue to dominate consumer electronic in the foreseeable future, opportunities exist for the United States and other industrial countries to capture an increasingly large part of the market. The JTEC panel also concluded that they found no insurmountable barriers that would prevent the United States from regaining a significant share of the consumer electronics market; in fact, there is ample evidence that the United States needs to aggressively pursue high-volume, low-cost electronic assembly, because it is a critical path leading to high-performance electronic systems.

ELECTRONIC PACKAGING

The term "electronic packaging" means the production and assembly of a great many types of increasingly tiny and complex electronic circuitry components central to the design of low-cost assembly of electronic products. Japan's manufacture of products like camcorders, palmcorders, handycams, VCRs, and cellular phones has simultaneously driven the miniaturization of electronic packaging and a corresponding advancement in assembly technologies. These popular consumer electronic products utilize a relatively large amount of analog circuitry, which has pushed the Japanese to develop cost-effective processes for assembling high-density miniaturized passive components. The use of "1005" packages (100 mm x 50 mm) and smaller formats requires both leading-edge surface mount process capabilities and ultrasmall component developments. That is, Japan's successes derive not only from production of advanced devices but also from the development of new equipment and processes to manufacture and emplace those devices.

As packaging technologies become essential to next-generation products, an increasing proportion of components and the equipment needed for production must be imported by the United States from Japan, or other Asian countries. The market for such electronic imports is expected to increase tenfold over today's market, representing an excess of $100 billion, by the year 2000 (JTEC Report 1995, p. 60).

Current trends in electronic packaging include decreasing feature size, lower costs, lighter weights, and fewer defects, accompanied by increased assembly speed, integration, and automation. Testing is a higher percentage of production costs and equipment is increasingly complex. It is also evident that there will be an increase in thermal and mechanical problems, as the density of the components increase and there is a corresponding reduction in the size of the final product.

In 1993, a study conducted by Microelectronics and Computer Technology Corporation and Sandia National Laboratory concluded that Japanese IC (integrated circuit) packaging and assembly were significantly superior in terms of miniaturization and cost effectiveness. It also concluded that the Japanese market share in consumer electronics had created the "product pull" required to justify large investments in manufacturing processes that, as a result, helped all other electronic segments, such as supercomputers and telecommunications (MCC/Sandia 1993).

Figure 8.1
Mass Production Strategy for Low-Cost Electronic Packaging Advances

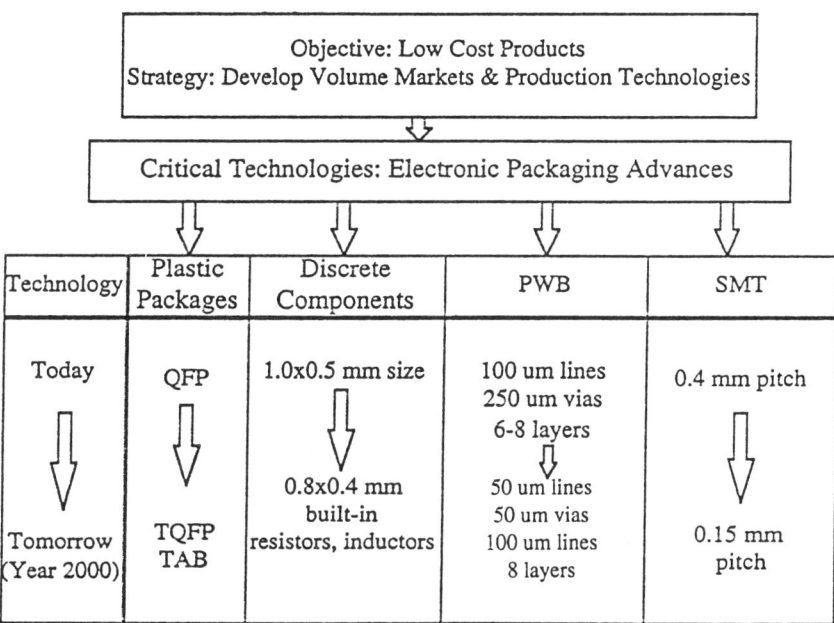

JAPAN'S ELECTRONIC PACKAGING STRATEGIES

Japan's current electronic packaging competitive advantage over the United States is a result of the breadth of its packaging expertise. Japan is investing in packaging technologies that are required to protect existing markets and in new technologies to grow next-generation electronics markets. Japan's electronics industry has greatly invested in the full range of packaging technologies, which include plastic and ceramic packages, multichip, passive component, printed wiring board (PWB), and surface mount technologies (SMT). To keep costs down, Japan continues to stress the incremental development of packaging technologies for mass production (Figure 8.1).

PRODUCTION TECHNOLOGIES

Major collaboration among industrial partners and between government and industry, as conclusively demonstrated in Japan, is required to sustain a competitive posture in high-volume, low-cost electronics business. Japanese industry as a whole has focused a massive amount of resources on the design and development of complex automated equipment, and the range of equipment development expertise within individual companies is tremendous. That expertise covers all the technological areas required to be self-sufficient and dominant in this field. The consensus in Japan is that equipment provides a major competitive advantage and that equipment development technology is mandatory in order to lead in the introduction of products. In a recent study done by Japanese Technological Evaluation Center (JTEC) for the Department of Commerce, the JTEC panel found that:

- Equipment is the key to advanced manufacturing; it must be an integral part of technology development.
- Investments required for automation to achieve precision assembly, manpower reduction, and agility must be balanced against requirements based on modular product design and modular assembly.
- Continuous improvements in existing processes avoid capital investments, retraining, and risks associated with the introduction of new technology.

Component miniaturization, cost reduction, reduced development cycle times, and improvements in reliability and quality require continued advancement in production and process technologies. Production requirements will include more affordable and environmentally safe materials; flexible and automated equipment linked to affordable manufacturing processes; cost effective and accurate testing; effective partnering with suppliers and enterprise teaming; continuous process improvements; and innovative, user-friendly designs. Increased demands for chip attach technologies will supplement current surface mount technology. Differences between technological alternatives will fade as technologies converge and hybrid electronic assemblies become commonplace in integrated systems.

The principal technologies that have provided competitive advantages include surface mount technology and flexible assembly technologies capable of responding to high-volume production with multiple product variations. While Japanese companies continue to do R&D in advanced process technology, it appears evident that surface mount technologies will continue to dominate consumer products into the next century. Mounting methods will become more sophisticated and include greater levels of chips and direct bonding of bumped chips. Mounting densities will increase to fifty components per square centimeter. Passive components are expected to reach their size limitation at 0.8mm x 0.4mm before they are integrated into modules. Pin pitches will be as low as 0.15mm. Low-cost resin board technologies will reach fifty micron lines and 50 micron vias with eight layers.

In the consumer electronics of the future, it will be increasingly difficult to separate integrated circuits, electronics packages, and flat panel displays.

While continuous improvements can be expected in materials, equipment, and design tools, it will be the flexible automated and adaptive manufacturing processes that will provide a competitive advantage. It is production technology that is making Japan the leader in high-volume, low-cost electronics, and it appears evident that this same strength will continue into the future.

THE SUPPORTING TECHNOLOGY AND MANUFACTURING INFRASTRUCTURE

After four decades of development in electronics and manufacturing technologies, Japan's electronic companies are leaders in the development, management, and support of complex, low-cost packaging and assembly technologies used in the production of a broad range of consumer electronic products. Japan's technology and manufacturing infrastructure is an integral part of its microelectronics industry's success (JTEC Report, 1995 p.35).

The infrastructure that supports Japan's leadership in consumer electronics (JTEC Report 1995, p. 38) is characterized by:

- Firms that appoint a greater number of technically trained managers to head their companies than do U.S. firms.
- Corporate enterprises that are structured and managed, to operate effectively in the global marketplace.
- Firms that have separated research on production technologies from other research and development activities.
- Enterprises that invest heavily in the development of production automation technology.
- A patent system that functions to facilitate industry-wide transfer of technology rather that to protect intellectual property rights.
- Industrial research that includes support through local municipal industrial research institutes.
- Firms that are constantly in pursuit of new technical knowledge and effective transfer of technology from global sources.
- The primary sustainable competitive advantage being people; information technology can be employed to support continuous learning, but it is not, however, a substitute for the kind of training and personal communication evidenced in Japan.
- A borderless manufacturing world that is a continuing reality that justifies further investments in transportable manufacturing-enabling technologies.

The existing infrastructure supports movement into advanced technologies and products. This is particularly evident in the electronics industries. Production development investment focuses attention and resources on manufacturing advancements that ensure the rapid introduction of new, high-quality products at low costs. Without advanced equipment capabilities, it would take much longer for new component technologies to become part of next-generation product designs. The strategic importance of manufacturing is emphasized in the education of the workforce and in the priority that management gives to continuously improving the process in order to more rapidly and efficiently manufacture complex products.

THE JAPANESE GOVERNMENT—ITS ROLE IN THE HIGH-TECHNOLOGY INDUSTRY

The Japanese government's support for high-technology is at least four decades old (JTEC Report 1995, p. 36). What is now known as the Ministry of International Trade and Industry (MITI) is the reorganization of the previous Agency of Industrial Science and Technology. MITI serves to guide Japan's technological developments.

The passing of two very crucial laws, the Machinery Industry Law in 1956 and the Electronics Industry Law in 1957—to promote the experimental research and initial production and to promote industrial rationalization of machine tools and electronic technologies—was influential in Japan's gaining its present status in the high-tech marketplace. Under these laws, subsidies for technology R&D were provided, along with special loans and tax incentives for firms that developed or used advanced production technologies. As a result, by 1980 Japan had become the world leader in robots for assembly and machine tools for component production (JTEC Report 1995, p. 36).

Approximately one-fifth of technology R&D comes from government sources in Japan. (Note that approximately one half of technology R&D spending, in the United States, comes from government sources.) The Japanese Government R&D policy supports the belief that basic research should be carried out by national laboratories and universities, and government R&D funds are allocated to this research. However, the government has attempted to stimulate industrial R&D in areas that will be advantageous in the future to the nation's industrial technology base. The government has primarily served as the supplier of strategic leadership and as the *catalyst* for private R&D investment. Capital markets in Japan support longer-term investments in technology, training, and other activities that ameliorate a firm's long-term growth potential, as compared with U.S. capital markets (DOC/TA 1993).

The Japanese government's policy of providing leadership and promoting private investment for advanced technology development has demanded extensive communication between industry and government in determining R&D priorities. The lack of antitrust enforcement has resulted in simplifying the cooperation and coordination of industry research activities. In many ways, the government's policies serve to support and maintain the interdependence within Japan's high-technology industries (JTEC Report, 1995, p. 37). The lesson from Japan is that the teaming of private industry and government is a primary success factor.

THE STRATEGIC ADVANTAGE—CREATING A VISION OF THE FUTURE

Behind Japan's leadership in the electronic packaging industry lies its strategic advantage in electronics production and process technologies. It is believed that Japanese competitors could be leading U.S. firms by as much as a decade in some electronic process technologies. Process technology improvements have allowed quality improvements and cost reductions in end products. Japan's continuous perfection of its electronics manufacturing systems has enabled it to take market leadership away from technology innovators in the United States.

In assessing Japanese critical factors, it is important to understand not only what technology Japan is developing, but also how firms pursue their objectives. In the study, JTEC also found that Japanese companies:

- Seek to identify customer needs as the basis for developing next-generation products that establish the roadmaps for technological development.
- Make long term commitments to component and equipment development that support future product innovations.
- Effectively utilize existing investments in the established supplier base and existing technological infrastructure; investments in new technologies are introduced only when competitive challenges require them.

The changes taking place in consumer electronics have significant implications for the future. Companies in the high-volume electronics business are on a steep learning curve that is providing continuous opportunities to fuse technologies to meet product objectives. This is most evident in the flat panel display technology that is merging traditional electronics with displays.

It appears that component vendors are moving toward supplying functional modules, and system integrators are becoming increasingly aware of the benefits of manufacturing components. Sony, for example, now manufactures about 65 percent of the key components of the compact disc player. There is evidence that companies such as Murata and Nippondenso are seeking increased independence through technology and component self-sufficiency. As vendors provide more of the subsystem integration and component costs increasingly dictate profits, the relationships between suppliers and end-product producers may change. At the time of the JTEC study, there was no indication of any lessening of the traditional supplier-customer interdependencies. Subcontractors contribute to new product development, and technical information is widely shared among vendors and end-product integrators. This organizational structure lends itself to effective concurrent development, shorter development cycle time, and lower life-cycle costs.

As the fusion of technologies increases, and electronic packages incorporating semiconductors, displays, and peripheral devices become increasingly integrated, the electronics industry may undergo more restructuring. It is clear that advanced technology and flexible manufacturing will not, by themselves, provide the advantage; nor will excellence in design. Organizations capable of quickly responding to change, led by visionary and capable management, will hold the essential competitive advantage.

Japan continues to be the world's benchmark for innovative consumer electronics, as well as displaying excellence in manufacturing and the quality of its workforce. Japan will continue as the leader in high-volume, low-cost electronic assembly, but there are opportunities for new product leadership. Some key factors to retain when planning to compete in electronic marketplace are:

- **Equipment** is the key to advanced manufacturing processes and therefore needs to be developed concurrently with technology.
- **Product cycle** time is reduced through concurrent development; systems and manufacturing engineering must play a dominant role.
- **Customer satisfaction** is driving innovative product design. The emphasis on simplicity is increasing, as a high rate of product change is confusing customers.
- Investments are almost totally dictated by **product pull**, however, long-term product planning permits "right-level" investments.
- **Continuous improvement and the pushing of existing technology** to its limits is preferred over the introduction of new technology, therefore it is advisable to fully exploit existing infrastructure and investments.
- **Define product requirements** and then concurrently develop technology and processes. Introducing the best technology implies the lowest cost.
- **Continuously transfer technology** through the constant movement and interaction of people.
- Maintain an **aggressive and knowledgeable management** with strong teaming. It is equally, if not more important, to have a management team that is capable of adjusting to managing through difficult periods as well as good times.

The issue that separates the U.S. from Japanese manufacturing in the electronic assembly and packaging arena is the will and not the means. The United States does have the means to "play" in this arena; if the United States does not "play," it will not win.

REFERENCES

Boulton, William R., Eugene S. Meieran and Rao R. Tummala. "Japan's Technology and Manufacturing Infrastructure," *The Japanese Technology Evaluation Center Panel Report on Electronic Manufacturing and Packaging in Japan*, 1995.

Microelectronics and Computer Technology Corporation and Sandia National Laboratory (MCC/Sandia). *Industrial Competitiveness in the Balance: A Net Technical Assessment of North American vs. Offshore Electronics Packaging Technology*, (U.S. Department of Energy Contract #AD-3474,) 1993.

Tummala, Rao R., and Michael Precht. "Japan's Electronic Packaging". *The Japanese Technology Evaluation Center Panel Report on Electronic Manufacturing and Packaging in Japan*, 1995.

U.S. Department of Commerce Technology Administration (DOC/TA). *Report of the U.S.–Japan Technology Transfer Joint Study Panel*, PB93-18292, 1993.

Part III

INTERNATIONAL TRADE, TECHNICAL STANDARDS, AND BARRIERS IN JAPAN

9

The Evolution of Technical Standards and Trade in a Changing World Economy

Stanley I. Warshaw

INTRODUCTION

Technical standardization needs to be viewed today in an increasingly competitive global environment. In today's world the United States can no longer impose itself, and its technical standards, as easily as it did from the postwar period until the 1980s. There are many countries with capabilities for providing quality products. There is a number of economic alliances forming among corporations that lead to joint ventures, subcontracting, dispersed production, etc. Foreign investments are expanding intra-firm trade across national borders. Acutely aware of the relationship between national economic well being and performance in the international marketplace, many nations are adopting trade policies to ensure their competitive position in the global market.

THE EFFECTS OF ECONOMIC INTEGRATION

The economic integration of regions such as the European Union and North America is serving to enhance each region's global competitiveness and liberalize international trade. Trade liberalization has its modern origins in the General Agreement on Tariffs and Trade (GATT) adopted in 1948 to achieve free trade. The GATT identified standards as a key issue related to trade barriers as early as 1967 (the Kennedy Round of negotiations). The resultant Multilateral Trade Negotiations Code on Technical Barriers to Trade, known as the "Standards Code," came into effect on January 1, 1980. It specified that international standards are to be used wherever possible in order to facilitate international trade. International standards are playing an increasingly important role today in both international trade and domestic commerce.

In our hemisphere, thirty-four nations have recently subscribed to the establishment of a Free Trade Area of the Americas (FTAA). Work on the FTAA began in 1995, and the first ministerial meeting took place on June 30, 1995, in Denver, Colorado. This ministerial meeting will establish an action plan for moving toward an FTAA. Ministers will agree on subject areas for immediate and longer term action. The United States has proposed standards and conformity assessment matters, for immediate action, on its agenda.

TECHNICAL TRADE BARRIERS

The United States efforts in the FTAA for standards related matters will encourage implementation of the principles of the Agreement on Technical Barriers to Trade (TBT). These principles were agreed to in 1994 by 124 nations in the Uruguay Round to the GATT. The TBT is designed to eliminate the use of standards related measures as barriers to trade and to encourage the transparent development of international standards and conformity assessment systems without compromising national, safety, health, or security measures. The TBT is based on the principle of "national treatment"—simply stated, the TBT provides that products or materials originating in the territory of another signatory shall be accorded treatment that is no less favorable than that accorded like products of national origin or like products originating in another country. The TBT agreement does not provide for an automatic right to gain recognition under another country's laboratory accreditation, inspection, or quality system registration scheme. It does encourage signatories to negotiate mutual recognition of the results of each other's conformity assessment procedures, known as Mutual Recognition Agreements (MRA).

A major difference in the TBT mandated by the Uruguay Round revision is that instead of a limited number (forty) of signatories to the GATT subscribing to the TBT provisions, now all the members of the successor organization to the GATT, i.e., the World Trade Organization (WTO), are obligated by the provisions in the TBT. As 1995, eighty-six nations are members of the WTO. It is expected that all thirty-four nations that participated in the Summit of the Americas last December will join the WTO.

The United States' efforts in implementation of the North American Free Trade Agreement (NAFTA) with our Canadian and Mexican partners support the TBT principles; and the possible entry of Chile into the NAFTA will not alter this position. NAFTA strengthened the standards-related obligations in the United States–Canada Free Trade Agreement (CFTA) and even exceeded the disciplines of the TBT by establishing a trilateral governmental committee on "Standards-Related Matters (SRM)." Specific product sectors for standards harmonization efforts are being addressed through the creation of subcommittees on land transportation, telecommunications, textile and apparel

The Evolution of Technical Standards

goods, and automotive. The MERCOSUR agreement among Argentina, Brazil, Paraguay, and Uruguay includes addressing standards related issues. Other bilateral and multilateral agreements with standards related provisions exist between and among Latin American countries.

TECHNICAL STANDARDS

Latin America

In Latin America, nations like Brazil, Chile, and Argentina are looking to international standards and conformity assessment practices as a means to improve their opportunities for exports. The demands by business in the international marketplace have prompted industry in these nations to seek certification to international standards of quality assurance (ISO 9000); Argentina has thirty companies so certified, Brazil over 450 quality assurance certifications, and Chile over 10. All have many applications for certification to the ISO 9000 standards in process. The demand for manufacturers of all goods to be certified to the ISO 9000 quality assurance standards has been growing rapidly in every nation; and the demand in two or three more years for industry to be certified to the forthcoming ISO 14000 environmental management standards is anticipated to be even more profound. Failure to understand the implications of these international standards for quality assurance or environmental management will have serious consequences for anyone engaged in international trade.

Europe

The European Union's efforts to integrate its economy and to expand membership in the Union to other European nations is "centered" today on the use of standards to provide definition for their mandated requirements for goods and materials entering into commerce. The mandated requirements for various goods and services in the European Union are characterized as "essential requirements" in the directives issued by the Commission of the European Union. The "new approach" that was adopted in 1985 to harmonize standards as opposed to harmonizing regulations throughout the European Union presents an attractive opportunity for increased trade by third countries, such as the United States, since national requirements among member nations cannot differ. However, since the entire or major product's required conformance measures by third parties must be conducted within the geographical bounds of the European Union, it places U.S. industry at a competitive disadvantage.

The European nations have always been very active players in the development of international standards through such entities as the

International Standards Organization (ISO) and its sister organization, the International Electrotechnical Commission (IEC), as well as organizations within the United Nations, such as the United Nations Economic Commission for Europe, the International Telecommunications Union (ITU), the World Health Organization (WHO), et al. The United States has previously not been an active participant in the development of international standards. Now, due in large part to the European Union's efforts to strengthen its regional development of standards, the U.S. is becoming much more active in the international standards arena. As a defensive measure, nations outside the European Union encourage the Europeans to adopt international standards instead of their European regional standards; i.e., those issued by the European Committee for Standardization (CEN), the European Committee for Electrotechnical Standardization (CENELEC), and the European Telecommunications Standards Institute (ETSI). The Europeans will have much greater leverage to promote their regional standards in the international arena as the Union expands to include additional member nations from Scandinavia and Central Europe. This will cause an even greater demand for increased U.S. participation in the development of international standards by the U.S. and other nations outside the Union.

United States

In April 1994, the U.S. Department of Commerce's National Institute for Standards and Technology (NIST) announced the establishment of the National Voluntary Conformity Assessment Systems Evaluation program, called NVCASE. It is a program that provides for U.S. government assurance of a given accreditation or conformity assessment program, conforming to international standards or guides for satisfying the regulated market sectors in another nation. The impetus for its development came about in 1991, as announced in a joint communique between the then secretary of commerce, Robert Mossbacher, and his counterpart in the Commission of the European Union, Vice-President Martin Bangemann. Negotiations for MRAs with the European Union require a governmental assurance of a conformity assessment body's competence. In the United States most conformity assessment bodies are in the private sector and are utilized by federal, state, and local regulatory agencies. NVCASE is designed to apply to any instance where a governmental assurance is required to gain acceptance of U.S. conducted conformity measures to foreign product regulations. NVCASE is not applicable to the unregulated market sector, and it does not address the U.S. market, regulated or unregulated. NIST has already achieved recognition of the laboratories it accredits through MRAs with several countries, but NVCASE addresses all aspects of conformity assessment, such as product certification, quality assurance, type approval, and so forth.

Asia

Last month I was privileged as the only invited participant from a Western nation to observe a meeting of the Newly Independent State's (NIS) Intergovernmental Council for Standardization, Certification and Meteorology (IGC). The IGC is comprised of the principal standards officials of the twelve NIS, namely Armenia, Azerbaijan, Belarus, Georgia, Kazakhstan, Kyrgyzstan, Moldovia, Russia, Tajikistan, the Ukraine, and Uzbekistan. The IGC operates under an NIS Economic Council, which is comprised of the Ministries of Foreign Economic Affairs of each state. These are the same ministers that co-chair with the secretary of commerce and the individual State Joint Business Development Committees in which the U.S. review nontariff trade barriers resulting from standards related matters. Clearly, the former Soviet Union (FSU) nations already recognize the need to harmonize standards and conformity assessment practices among their member states and internationally. While these emerging market economies are not major players in international trade today, their abundance of natural resources and well educated technologists strongly suggests that they will be a major force in future international trade. U.S. trade with Russia has increased over 300 percent this year from last year, and American investment in Russia is currently approximating one-half billion dollars monthly.

In the Asia-Pacific Economic Cooperation (APEC), representatives of Australia, Brunei Darussalam, Canada, Chile, the People's Republic of China, Hong Kong, Indonesia, Japan, the Republic of Korea, Malaysia, Mexico, New Zealand, Singapore, Chinese Taipei, Thailand, and the United States serve as a Sub-Committee on Standards and Conformance (SCSC). They are looking to align their national standards with international standards and conformance practices. Current efforts are focused on air conditioners, television sets, refrigerators, food labeling, rubber gloves and rubber condoms, and unplasticized plastic pipes and fittings. Contributions to these developments are also being obtained from such private entities as the Asia-Pacific Legal Metrology Forum (APLMF), Asia-Pacific Meteorology Program (APMP), Pacific Accreditation Cooperation (PAC), and the Pacific Area Standards Congress (PASC).

Japan

Japan is well known to have used standards as nontariff trade barriers for some years. The most infamous case was that of the Japanese standard for metal baseball bats and the requirement for lot inspection at the port of entry into Japan. The controversy between Japan and the United States was aired on both nations' television networks during the late seventies and early eighties. This dispute was also brought before the GATT in Geneva before a satisfactory resolution could be achieved. As a result of the action brought before the GATT by the United States, the Japanese agreed in 1983 to accept the results of the testing of products by laboratories outside Japan.

The Japanese government through the Ministry of International Trade and Industry (MITI) and its Agency for Industrial Science and Technology (AIST) operates the Japanese Industrial Standards Committee (JISC). JISC was established by law as an advisory organization to the government, on the use of Japanese Industrial Standards (JIS) and the JIS mark of conformance. Trade associations and professional societies that develop standards outside the government-approved JIS standards usually coordinate their efforts with AIST and other ministries. There are little, if any, redundant or overlapping standards in the Japanese system. The Japanese Standards Association (JSA) is responsible for publishing and distributing all JIS documents.

Since 1985 the Japanese government, as part of its Action Program for Improved Market Access, has promoted access to drafts of Japanese standards, whether within the JIS system or not. At the urging of U.S. industry, both the U.S. embassy and MITI have had to intervene on several occasions to remind Japanese standards developers of this obligation, and the fact that Japan became a signatory to the GATT in 1980. The Japan External Trade Organization (JETRO) in MITI and the Ministry of Foreign Affairs operate the required "inquiry point" under the GATT. JETRO's offices operate around the world as intelligence gathering points for MITI on technical standardization activities in various nations and international organizations. Clearly, the Japanese system of managing standards activities at the Federal government level differs markedly from that of our own private sector-led pluralistic approach to both standards development and conformity assessment practices.

CONCLUSION

In summary, we have some good principles contained within the TBT for reducing standards related trade barriers and the incurred costs resulting from restrictions to entering any given market. We have many forums to communicate with others on standards and conformity-assessment matters, and, more important, we have direct involvement by the private sector, from the United States, in our government led negotiations in order to ensure our ability to operate competitively throughout the world. The two questions that I perceive as important to consider is whether industry can provide the resources required to be an effective participant in international standardization, and whether alternative approaches to current United States standards development and conformity assessment activities are needed to achieve international business objectives. These are not simple questions and the answers will vary for different product sectors. I urge that these questions and their implications be considered as business strategy is developed for Japan and elsewhere.

10

Technical Regulations and Trade: New Developments and the Asia Pacific Economic Cooperation Forum

John Sullivan Wilson

INTRODUCTION

Technical regulations and national systems for testing, certification, and laboratory accreditation are an increasingly important part of industrial production and trade (Wilson, 1995). Efficient standards systems can accomplish several important goals. Standards can facilitate technology diffusion, as well as support economies of scale through improved production techniques (Farrell, 1987). The benefits of open and transparent standards systems also involve the assured compatibility of key components in national infrastructure, including telecommunications and computer networks, for example (Besen, 1986). Technical regulations can also support public welfare by promoting quality assurance systems and health, safety, and environmental goals. Finally, the development of modern, efficient standards and conformity assessment procedures are central to raising product quality and the expansion of global export markets.

The Asia-Pacific Economic Cooperation (APEC) forum, recognizing the benefits of a common approach to standards and conformance, has included this work as part of its trade facilitation agenda. The standards and conformity assessment agenda of APEC represents one of the most critical components of the drive to achieve open trade in the region by 2020. APEC may be particularly well suited to promoting economic cooperation in the Asia-Pacific region through standards initiatives. These are issues of prime importance to both the developed and developing members of APEC for many reasons. As global tariff rates have fallen and nontariff barriers to trade are reduced, technical barriers now represent some of the most serious obstacles to regional trade expansion. Approximately 15 percent of all notifications to the GATT of nontariff barriers in the past decade, for example, involve some form of technical regulation. APEC members recognize the need to provide a stronger regional framework to mitigate against a continued rise in these disputes. Moreover, the world trading system, especially firms in rapidly expanding regions of Asia, will also confront new requirements in environmental standards introduced by the International Standards Organization (ISO) in draft form, in 1995. In addition, as a greater share of economic activity is developing within APEC's members, driven by private firms, building a modern standards infrastructure to support quality, safety, and reliability has taken on heightened importance to many APEC members.

There are social welfare benefits associated with standards. The work required of manufacturers to achieve and demonstrate conformity, however, has the potential to present costly new barriers to international commerce. APEC's work to promote regional dialogue, therefore, represents a unique opportunity to address these issues as part of broader trade facilitation efforts. Opaque standards setting rules adopted to favor inefficient domestic producers over foreign competitors can retard technological advance, which is of critical importance to all APEC membership.[1] Technical barriers to trade embedded in discriminatory certification requirements present some of the most serious problems for firms operating in global markets today. These go beyond instances where there are differences between national standards or cases where countries refuse to accept any foreign test data. Some of the less well understood, but serious, problems in world trade involve: (1) opaque testing and certification requirements often set at higher standards for imports; (2) costly, confusing, and discriminatory product labeling rules; (3) manipulation of domestic laboratory accreditation regimes to block imports; and (4) mandatory compliance with unnecessary and unduly burdensome quality system registration schemes.

This chapter will review new international initiatives to streamline and reform standards and conformity assessment systems. It will focus on the possible contributions of APEC to regional and multilateral trade liberalization through this international work. The material addresses mandatory government product standards and technical regulations. The chapter does not discuss issues associated with sanitary and phytosanitary (SPS) standards[2] specific to the protection of human health and regulation of trade in agricultural products. As the prospects for success in APEC initiatives in product standards will be affected by developments at the multilateral level, the 1994 Technical Barriers to Trade (TBT) Agreement will be outlined. Key provisions of the agreement will be discussed, especially as they relate to strengthening of disciplines in the Tokyo Round TBT code.

One of the most important parts of APEC's 1995 work program involves regional negotiations on Mutual Recognition Agreements (MRAs) in conformity assessment. The current European Union (EU)-U.S. negotiations on an MRA will undoubtedly contribute to the substance and models developed by APEC. The paper will examine the relationships between these negotiations and conclude with comment on how APEC can contribute to multilateral and regional efforts at harmonizing conformity systems, as part of an action agenda for standards to the year 2020.

THE URUGUAY ROUND TBT AGREEMENT

The Uruguay Round concluded in 1994 with the signing of a world trade agreement. Significant progress was made in advancing the goal of reducing barriers to trade, including acceptance by national governments of an Agreement on Technical Barriers to Trade (TBT) (*Agreement on Technical Barriers to Trade*, 1993). The agreement addresses "product characteristics or their related processes and production methods," as reflected in mandatory administrative provisions of governments in technical regulations. The TBT agreement and related standards provisions of the new GATT agreement have important implications for APEC member governments, especially those nations not currently signatories to the Tokyo Round Code. There is a great deal of work necessary to conform national laws and procedures to international norms, or even simply reform of standards development. Policies and institutions, especially in the less developed areas of the Asia-Pacific, will need to modernize to meet provisions of the UR on transparency and removal of trade-distorting technical regulations.

The TBT provisions not only remove many existing technical barriers to trade reflected in the preparation, adoption, and application of national standards, but also will act to prevent the creation of new barriers, particularly in discriminatory conformity assurance systems. The following represents a summary of the key elements of the Agreement, as contrast to the existing Standards Code negotiated in the Tokyo Round, and outlines areas of uncertainty in TBT implementation and impact on trade.[3]

Membership and Scope

The most important progress made in the Uruguay Round TBT agreement is directly related to the expanded scope and coverage of international disciplines on technical regulations, as they affect trade. This includes both the increase in the number of countries bound by the obligations in the new agreement, and the fact that TBT rules, for the first time, will cover *processes and production methods*, as well as manufactured products. This expansion is an extremely important move toward strengthened international discipline. As of November 1993, there were forty-six signatories to the Tokyo Round TBT code. Most of these members are the industrialized nations of the European Union, with the United States and selected other Asian and Latin American countries represented. The signatories to the Uruguay Round Agreement and new TBT Agreement include sixty-eight additional nations. Many of these include the most rapidly developing nations of Asia and Latin America, including APEC member China. Other members of APEC, which have just recently signed the Tokyo Round Code and are therefore new to international disciplines, include Indonesia. Based on 1991 data, new signatories of the Uruguay Round TBT Agreement represent an expansion of approximately $182 billion in global imports subject to international discipline. This is a 17.5 percent[4] increase over total imports covered under the Tokyo Round Standards Code.

The extension of rules and procedures on standards and conformity assessment is an important part of strengthening the multilateral trading system. The TBT code helps support progress toward global market liberalization worldwide. It furthers the process of binding the developing nations with their industrialized trading partners in an area of increasing importance to the trade system. This is perhaps particularly important in the APEC region, as the economies of East Asia, excluding Japan, are expected to grow 6.2 percent per year from 1993–2000 and constitute 27.9 percent of global GDP by the year 2003.[5]

In addition, the agreement subjects processes and production methods to the same rules applied, under the Tokyo Round Code, to manufactured goods. This was a significant achievement executed through changes in the terms "standard" and "technical regulation" in Annex 1 of the agreement. The expansion of coverage has the potential to mitigate against measures to block imports of products based on the way in which they are produced. There have been several high profile trade disputes in this area in recent years. Most notably, the U.S.–E.U. trade in agricultural products and the Community's Third Country Meat Directive and Beef Hormones Directive case. At a minimum, a transparent, multilateral framework through the TBT Agreement now exists to address these procedures, as they relate to international trade.

Conformity Assessment

National rules and regulations for the oversight of product testing and certification represent one the fastest growing segments of most industrialized nations' standards systems. Although most of the attention in trade policy discussions has involved benefits associated with harmonization to international standards, expansion of national regulations on conformity to standards are where costs to manufacturers and exporters, especially those in the emerging economies of APEC that will modernize industrial production facilities over the next several decades, are likely to grow. The rapid increase in the number of technologically advanced testing laboratories and national systems for laboratory accreditation in environmental emissions, for example, will contribute to growing complexity in these systems.

In the developing nations, rising incomes, continued industrialization, and increased exposure to export markets will undoubtedly contribute to expansion in testing and accreditation systems over the next decade. The rapid growth in registrations by firms in Asia to de facto mandatory ISO 9000 quality guidelines is one example of the continued proliferation of regulatory mandates in conformity assessment. Registrations to the ISO 9000 series in the Far East grew by more than 500 percent from June 1993 to June 1994. As part of the rapid expansion in private sector economic activity in Indonesia, for example, registrations by firms to ISO 9000 guidelines are rising sharply. These grew from several manufacturing plants in 1990, to over twenty-five by August 1994.

Other APEC member nations will continue to expand complex testing and certification infrastructures. In the United States, for example, the independent testing industry represented approximately $10 billion in revenue in 1993, with average annual growth rates of 13.5 percent from 1985–1992.[6] As third-party testing increases in frequency, so does growth in the number of accreditors of test laboratories and, most likely, expansion in regulatory guidelines for oversight of both laboratories and accreditors. These systems all impact the ability of exporters to service national markets. Manufacturers in developed and industrializing nations will be subject to the full range of conformity assessment steps in export markets. Third party firms are already displacing the use of manufacturers' self-declaration of conformity in a number of sectors; most notably capital goods (Locke, 1993). Protocols at the international level, therefore, to bring clarity and foster acceptance of these systems across national borders will be increasingly important to trade.

The Tokyo Round Code applied the principles of national treatment and non-discrimination only to product testing and certification programs. It did not address the full range of conformity assessment procedures. Articles five through nine of the Uruguay Round TBT Agreement contain the basic obligations of WTO signatories in regulated products and represents significant progress over the current code. The TBT Agreement extends the principle of national treatment and non-discrimination to not only testing and certification, but also registration, inspection, laboratory accreditation, and quality system registration programs. Extension of the coverage in the Uruguay Round to all forms of conformity assessment work, therefore, provides a framework to protect against the future use of these as barriers to trade.

The Agreement provides only a limited basis to encourage acceptance of the results of tests or laboratory accreditation across national borders. Article 6 of the TBT Code simply exhorts signatories to move toward harmonization of conformity assessment through mutual recognition of each others' procedures. These types of agreements have the potential to offer real benefits in reducing costs, inefficiencies, and barriers in international trade. This area of focus is especially relevant to APEC and where APEC leadership at the regional level may help foster progress at the multilateral level.

Non-Governmental Bodies and Local Government

Another important way in which the TBT Agreement represents progress involves the extension of rules to sub-national levels of government (regional, state, and local), as well as private standards organizations. Article three of the agreement calls for "reasonable measures" by members of the WTO to ensure compliance by these bodies with principles of national treatment, non-discrimination, and notification of standards preparation in advance of promulgation. The article further states that "members shall formulate and implement positive measures and mechanisms in support of observance of the provisions of Article two by other than central government bodies." Central government, therefore, will now be responsible for good faith implementation of the agreement and application of its principles at any level of government or within any private sector body involved in the standards system. The TBT Code of the Tokyo Round only bound central governments to these obligations.

The Code of Good Practice contained in Annex three of the TBT Agreement provides an additional foundation for executing the extension of rules to private standards bodies. The code outlines general principles for the preparation, adoption, and application of standards by nongovernmental organizations. These include national treatment, non-discrimination, publication and dissemination of work in progress, institution of a sixty day open comment period prior to adoption of standards, and refrains from applying standards that could serve as barriers to international trade. While adoption of the Code of Good Practice is voluntary and lacks enforcement mechanisms, it does outline for the first time in a multilateral agreement a common mode of operation for private standards bodies consistent with open trade.

Dispute Settlement

An extremely important point of progress in the Uruguay Round Agreement, lacking entirely in the disciplines of the current Technical Barriers to Trade Code, is a clear and binding framework for the adjudication of disputes. Potential disputes involving technical barriers to trade will now be settled as part of an integrated system under the WTO. Violations of, or noncompliance with TBT provisions found by dispute settlement panels of the WTO, will require action to either curtail trade distorting behavior or allow for retaliation. At a minimum, this will transform the manner in which the TBT principles are viewed by governments, introducing serious consideration of actions taken in standards and conformity assessment rules and procedures. There is the possibility, at least, that violations in these areas will involve internationally public consequences.

Implementation and Next Steps

The Uruguay Round Agreement made great progress in reducing the potential for use of standards as nontariff barriers to trade. There are, however, questions as to the implementation and lack of specificity in the agreement, which may limit its effectiveness. It is unclear, for example, how effective the agreement will be in affecting developments in environmental standards.[7] There is significant interest in developing new standards that link industrial production, trade, and the environment. Whether the WTO and TBT Agreement provide a suitable framework to foster the creation of least trade-distorting environmental standards remains questionable. Dialogue in APEC on these issues can be particularly useful in supporting these objectives. Moreover, APEC can play a role in providing a platform to ensure continued progress on harmonization of standards. There is only vague language in the TBT agreement that commits national governments to action in reaching this goal.

Another area of uncertainty with the new TBT Agreement centers on the difficult process of movement toward reciprocity in conformity assessment procedures. It is likely that problems of interpretation will arise in areas such as laboratory accreditation and quality systems registration in environmental management. It remains unclear, for example, how national governments will interpret Article six of the agreement, which requires "whenever possible, that the results of conformity assessment procedures in other Members are accepted, even when those procedures differ from their own, provided they are satisfied that those procedures offer an assurance of conformity." At best, the TBT Agreement offers only a general framework for multilateral discipline by promoting open, transparent, and non-discriminatory conformity assessment. Development by APEC of model MRAs to streamline acceptance of testing data, for example, could provide critical leadership for the international community in support of future liberalization at the WTO.

Progress in this area, in part led by APEC, is important to world trade. It is likely that discriminatory practices to block the importation of goods in the future may develop under the authority outlined in Article two, which allows for preparation, adoption, and application of technical regulations that are trade restrictive if they serve "national security requirements, or protection of human health, safety, animal or plant life or health or the environment." There is the danger, especially in the evolving systems of conformity assessment, such as laboratory accreditation and quality system registration, that governments will find a relatively easy route to new discrimination under these conditions. The

TBT agreement offers only weak encouragement for the MRAs on conformity assessment, one way to help forestall future disputes. The lack of specificity in the TBT Agreement on disciplines that encourage least trade distorting conformity assessment systems raises the importance of rapid progress in regional forums such as APEC. In particular, differences between APEC's framework for conformity assessment and the model MRAs developed in the U.S.–EU talks have long-term implications for future progress at the multilateral level and working of the new WTO.

CONFORMITY ASSESSMENT

A largely unexplored subject with major implications for trade involves procedures to determine conformity to standards. These include tests performed in laboratories to evaluate whether a product meets criteria set forth in a standard. It also may involve complex rules for the accreditation of laboratories as competent to perform these tests. Conformity assessment also encompasses national and international programs to evaluate manufacturer's quality systems. The most widespread of these quality programs are guidelines under the International Organization for Standardization (ISO) 9000 series, a set of quality assurance management procedures for manufacturers and service sector firms.[8] A well functioning national system to test and certify products serves public needs in ensuring health, safety, and environmental protection. It can also provide for cost effective and reliable procurement by government of goods and services (Locke,[9] 1993).

Conformity assessment involves one or more of the following four procedures: (1) manufacturer's self-declaration of conformity through internal testing and quality programs; (2) testing of parts, materials, and final products by independent laboratories; (3) formal certification by a third party that product conforms to particular standards, which often includes the granting of mark or label, such as the Underwriters Laboratory (UL) mark for electrical products; and (4) independent audit and approval of manufacturing quality systems, resulting in registration with a quality systems registrar. The program most recognized on an international level in this area is the ISO 9000 series.

The Expansion of National Systems

Conformity assessment regimes can provide value-added benefits in assuring the safety or reliability of products. They may also facilitate trade when mutual recognition of national systems is in place. This is most easily accomplished in regulated product sectors such as agricultural chemicals, electrical machinery, and pharmaceuticals, for example. Finally, manufacturers are demanding increased quality and independent assurance of conformity from suppliers. The increased use of strict inventory control and zero tolerance for defects in manufacturing production systems also heightens the importance of multiple checks on conformance through quality management system specifications.[10] This is especially true in sectors such as aviation, water systems treatment facilities, and nuclear power generation, for example.

Over the past several decades, especially in the industrialized nations of Europe, North America, and Asia, there has been an expansion in the complexity and duplication of systems to document conformity to standards (National Research Council, 1995). Government and private sector firms are increasingly involved in performing redundant: (a) testing, (b) certification, and (c) laboratory accreditation work. At the top of the system are a growing number of programs, primarily government operated, that seek to recognize the competence of laboratory accreditors to perform their duties.

There is only limited data and information available to compare conformity assessment on a cross-national basis. Most available information describes domestic testing and certification requirements of national governments. Anecdotal evidence indicates, however, that there is a movement toward: (1) increasing the number of third-party testing laboratories operating for-profit establishments in industrialized nations; (2) added requirements of manufacturers of quality registration to domestic or international standards, particularly to the ISO 9000 series; and (3) growth in demands that laboratories obtain multiple third-party accreditation of competence.

Systems in the highly developed APEC member nations, for example, continue to grow in complexity. There are several thousand independent testing laboratories in the United States alone. The revenue of these firms has risen sharply[11] since the mid-1980s. Although this may be, in part, due to restructuring in U.S. industry and a rise in sub-contracting of testing work to outside firms, it is also a response to increased national and international demands for third-party conformity assessment. Any increase in independent assessments will likely involve increased oversight activity. As demand for independent assessment of products generates demand for independent accreditation of assessors, complexity and cost propagates throughout the system. For example, in order to market their services widely, testing laboratories must often secure accreditation from multiple organizations with differing requirements. Laboratories operating in most APEC member nations are often required to perform duplicative tests and transmit data for the same sets of testing procedures to each accreditor. These services involve payment of annual fees, on-site audits, and provision of necessary documentation.

For example, there are more than thirty-one federal, thirty-two state and local, and forty-eight private accreditation programs in the United States (Breitenberg,[12] 1991.) One of the largest private accreditors is the American Association for Laboratory Accreditation (A2LA). Organizations such as A2LA, which offer national accreditation services, must also often secure acceptance by different federal agencies, states, cities, and firms operating in specific product sectors. The U.S. government, through the National Institute of Standards and Technology (NIST), also operates its own National Voluntary Laboratory Accreditation Program (NVLAP).

In one of the few studies performed in the United States on costs of conformity assessment, a group commissioned by the Environmental Protection Agency found that laboratories conducting similar environmental tests were subject to duplicative accreditation by multiple jurisdictions (Hankins et al., 1992). In 1991, the average costs per audit by accreditors totaled $4050, with labs subjected to reaudit every three years, on average. Another study of electrical safety-testing found that laboratories seeking accreditation must gain the acceptance of forty-three states; more than one hundred local jurisdictions; three building code organizations; federal agencies, including the Occupational Safety and Health Administration and NIST's NVLAP; as well as several large manufacturers (Breitenberg, 1991).

The New Nontariff Barriers to Trade

There are only limited data on the volume of global exports subject to conformity assessment regulations. The Commerce Department estimates,[13] however, that $300 billion of the $465 billion in U.S. merchandise exports, including those to APEC members, were subject to foreign technical requirements and standards in 1993. A total of $180 billion was subject to non-U.S. requirements or standards. Of the $110 billion in U.S. merchandise exports to Europe in 1993, $66 billion were subject to some form of EU required product certification. Approximately $30 billion required government-issued certificates, $25 billion manufacturer's self-certification, and $10 billion in exports were subject to private, third party certification.[14]

The addition of new layers of complexity and cost in commercial transactions is cause for concern. There is danger that continued proliferation of redundant conformity assessment requirements may present serious future problems in international trade. Duplicative requirements threaten to undermine the trade enhancing benefits of standards harmonization, by adding a layer of costly requirements for showing conformity to equivalent specifications in multiple export markets. What is necessary is a forceful effort at streamlining international systems in certification and quality regulations at the regional and multilateral level. Manufacturers servicing global markets should be able to obtain third-party certification of products and services and registration of quality systems only one time, in one market, and have this accepted globally.

Mutual Recognition Agreements (MRAs)

Mutual recognition by governments of test data, laboratory competence, and certification requirements represent the potential for increased trade. These agreements, as currently envisioned, would address third-party testing, inspection, and certification in sectors regulated by governments through product approval systems. An examination of the scope of the U.S.–EU talks on an MRA, launched in early 1994, illustrate the potential benefits of such agreements. These talks also offer insight into the possibilities for agreements in other regions, including an MRA within APEC.

The U.S.–EU negotiations on sectoral MRAs in conformity assessment were developed as a response to the on-going harmonization of standards in Europe. In 1990, the Community established a framework for mutual recognition within Europe under a "Global Approach" to testing and certification. This system builds on the framework approach to European harmonization initiated in 1985, which delegated formal standards setting to the private sector under broad "essential requirements" for regulated products. Conformity assessment is proven in Europe under general technical rules outlined by the commission. If third-party product approval is required by law, it is granted only by organizations "notified" to the commission by the member states. Products that meet conformity under these procedures are granted the CE mark and may circulate within Europe.

Although EU harmonization has made it possible for U.S. exporters to gain access to the entire European market through compliance to a single set of regulations, it has raised barriers to U.S. products at the conformity assessment level. The requirement that assessments be done by European "notified bodies" sharply raises the costs to U.S. manufacturers of testing and certification. In order to facilitate EU–U.S. trade, negotiations are in process to outline rules under which U.S. manufacturers would be able to obtain certification to EU requirements through a U.S. facility (through establishment of a U.S. notified body) in the United States. A European manufacturer would gain greater access to U.S. certification, including the possibility of conducting laboratory tests within Europe for regulated products.[15] MRAs are under discussion in ten industrial sectors: information technology and telecommunications products attached to public networks, medical devices, electrical safety, pharmaceuticals, pressure equipment, chemicals, food additives, pesticides, lawn mowers and central office telecommunications equipment.

There are a number of benefits associated with MRAs in these regulated sectors. Manufacturers would be able to obtain required national certificates at the location of production, rather than covering higher costs of certification offshore. This would allow direct shipping of products from point of production to final sale between Europe and the United States. In addition, MRAs may forestall the creation of added layers of new regulation in these sectors by increasing confidence in data and information developed in Europe and the United States. This would likely be the case, for example, with an MRA in the medical device industry.[16]

There are many obstacles to successful completion of these talks. The United States accepted the EU proposal to negotiate based on a framework for MRAs in the full range of conformity steps: acceptance of test data, laboratory accreditation, and final product certification within each industry sector under discussion. This has proven to be an extremely time intensive and costly process. As of May 1995, European negotiators have suggested that they will conclude MRAs only if all sectors under discussion are included in an agreement. Moreover, there are serious problems in that U.S. and European systems differ greatly in structure and operation. The United States relies primarily on voluntary, self-certification procedures, leading to government acceptance of conformance. The EU system centers on third-party conformity assessment in regulated sectors. Most important, Europe relies on "notified bodies," which provide oversight of laboratory accreditation and testing organizations for domestic and foreign products. Firms must be approved by a European notified body to perform testing and certification to EU directives. No foreign firms are able to gain this approval and may only operate as subcontractors to European-based companies. This system, which relies on third party services and blocks foreign competition, represents a clear barrier to international trade (Wilson, 1995, pp. 132–134).

The MRA talks between Europe and the United States offer one model of how to structure dialogue on conformity assessment. There may be other, more appropriate methods of providing a framework for regional harmonization. The APEC work agenda on standards and conformance can play an important role in providing alternative models to the ones being developed in the U.S.–EU dialogue. The following section outlines next steps for APEC on an MRA and other suggested priorities in standards to facilitate economic cooperation and trade in the region.

APEC'S STANDARDS DIALOGUE

The 1994 ministerial meeting of the Asia Pacific Economic Cooperation (APEC) in Bogor, Indonesia, officially launched a campaign to achieve open trade in the region by 2020. The Bogor Declaration proposed a set of initial steps to be taken to reach this goal. First, a commitment was made to accelerate the implementation of the Uruguay Round by the APEC members. Second, encouragement was given for individual APEC members towards further unilateral trade liberalization. Third, a standstill commitment was made to avoid policies that might serve to increase trade barriers. Finally, a series of trade facilitation work programs were initiated. These include work on an Asia

Pacific Investment Code, harmonization of custom procedures, and dialogue on standards and conformity assessment reform in the region.

Japan chaired APEC for 1995. The group is discussed an action agenda in standards for adoption by the group at the Osaka Leaders Meeting in November 1995. The Standards and Conformance Subcommittee (SCSC), led by Japan, was responsible for drafting this agenda. What are the most important priorities for APEC? They center on three objectives. The first involves implementation of the TBT Agreement of the Uruguay Round. A strong foundation for reform at the multilateral level will provide APEC with the basis for closer understanding and cooperation at the regional level. A major effort in APEC to provide information, training, and technical assistance on TBT implementation is essential. The second part of APEC's strategy to 2020 should include assistance in building modern infrastructure necessary to support reform. Technical assistance is also a necessary condition to support future work on a third priority initiative for APEC—regional MRAs in conformity assessment in regulated product sectors.

TBT Implementation

Providing a strong basis for cooperation within APEC to ensure implementation of the TBT provisions is essential to trade. This should be a primary focus of APEC's action agenda in standards. Work is underway sponsored by the SCSC to outline a regional implementation program. As a simple first step, the sub-committee should sponsor regional seminars to provide information and assistance on obligations of the TBT Agreement, methods of gaining assistance from the WTO on implementation, and other sources of aid in implementing the agreement. Programs should be conducted under the joint leadership of both developed and developing member nations. Seminars could be conducted soon in several locations in Asia in cooperation with the WTO. Australia, New Zealand, Japan, and Indonesia have already taken the lead in early work on standards and conformance in APEC and are logical candidates to coordinate this work with the WTO. Undoubtedly, this will require funding to support TBT implementation. It need not, however, be a costly project and should be supported by the wealthiest APEC members, perhaps with some aid from private sources in Asia.

Development of Technical Infrastructure

There are an extremely limited number of infrastructure projects funded by the World Bank, UN Development Program, Asian Development Bank, or other institutions that focus on standards work. One estimate of the annual expenditures on all technical assistance for industrial and agricultural standards projects by these groups totals approximately $180–220 million.[17] Funding is typically provided for projects such as upgrading laboratories to international standards, dissemination of information on international standards, quality, and production techniques in industry, and improving national capacities in metrology, for example. The International Standards Organization (ISO), the multilateral group that works to support global harmonization of standards, also supports standards systems in developing nations, primarily through training and education programs.

On an individual basis, industrialized nations also support standards systems and infrastructure needs in developing countries. Most technical assistance programs are tied to export promotion activities. Japan, for example, provides training courses for developing nations through the Japanese Standards Association (JSA) on standardization, conformity assessment, and quality control. There are also Japanese experts posted overseas to assist public and private officials with standards work in Saudi Arabia, Argentina, Chile, Malaysia, Thailand, and the Philippines. This assistance, as with national programs elsewhere, is provided with the expectation that the adoption of Japanese standards will be fostered, enhancing the prospects for sale of Japanese products.

In the United States, the Department of Commerce, through NIST, leads government assistance on standards, laboratory accreditation, and other conformity assessment measures. To date, there are resident U.S. government officials offering standards assistance to the Saudi Arabia Standards Organization (SASO), along with representation to standards officials in Europe through the U.S. Mission to the EU, in Brussels. NIST is planning an expansion of overseas technical assistance program to other countries. The agency would place resident experts in several member nations of APEC, including Mexico, Japan, South Korea, and China. The program will be coordinated with the U.S. Foreign and Commercial Service and industry to promote the dissemination of U.S. technology in standards and conformity assessment systems. A total of approximately $4.5 million would be spent on these activities.[18] The U.S. expert residents overseas will sponsor training seminars, as well as assist in managing visits by foreign experts for training in the United States. As part of these efforts, a new database for international standards projects is planned. The system would support the exchange of draft standards and dissemination of new standards through computer links with overseas partners.

Private sector technical assistance in the United States is largely executed on an ad-hoc, case-by-case basis. Most of work is provided with the assistance of the American National Standards Institute (ANSI). This private sector organization coordinates U.S. industry response to requests for assistance and conducts technical training programs, seminars, workshops, and other events. In July 1994, ANSI announced plans to sponsor a workshop to consider the establishment of a government-industry program to coordinate U.S. assistance in metrology, standards, testing, and quality (MSTQ). The proposed program would create a centralized operation to support delegations of technical experts for overseas assignments, develop training courses in MSTQ, and establish a permanent clearinghouse for MSTQ assistance.

A new program sponsored by APEC, on technical assistance, would need to leverage national government and private sector efforts, such as the ones outlined above. It is likely, however, that any effort of APEC would need to go beyond the type of assistance provided by individual governments in trade-related standards assistance programs. Similar levels of technical proficiency and competence among APEC nations is necessary for the success of any future dialogue on MRAs, for example. Funding by APEC could be administered through a regional organization, such as the Asian Development Bank, that would also monitor progress on both infrastructure development and the installation of systems that promote open, transparent, and harmonized policies. Any program of technical assistance should focus on upgrading the infrastructure in developing nations, including work in sectors noted by the Eminent Persons Group to APEC Ministers in its October 1993 report.[19] These include: compatibility standards for telecommunications networks, standardizing customs documentation procedures, and standards related to air traffic control systems.

MRAs in Conformity Assessment

The APEC agenda on conformity assessment is the most promising and important part of the forum's trade facilitation program. What are the prospects of agreements on a regional level in APEC? What are the lessons, if any, from the U.S.–EU talks on an MRA for APEC? The dialogue between Europe and the United States on MRAs is structured on an overly complex and unworkable scale. It should not be adopted by APEC. There are U.S.–EU negotiations in ten sectors, with large teams of private sector advisors and government officials involved in talks to achieve full acceptance of conformity procedures. The negotiations involve a model that has been structured to address all levels of conformity assessment—(1) recognition by government of accreditors, (2) laboratory accreditation work, and (3) issues affecting the testing facilities performing specific certifications. The Europeans, in particular, are interested in securing the goal of "complete market access," covering any regulatory, tort, or contractual conformity assessment procedure (Ludolph, 1994). There are clearly tremendous difficulties in such a highly detailed first approach to streamlining conformity assessment. APEC should develop its own model, suited to a more limited set of MRAs in the region in one or two industry sectors. Such a model has been proposed by the United States and other APEC members for discussion in November in Jakarta.

There are significant differences among APEC nations, by level of industrialization, nature of government control over the standards and certification systems, and state of infrastructure capacities. Certainly, a more measured approach for APEC on conformity assessment than that underway between the U.S. and EU is needed. MRAs in all regulated product sectors will be a long-term work project for APEC. This goal will not be achieved over the next fifteen years. APEC's agenda should focus first on ensuring the fundamental building blocks of MRAs are in place and harmonized in APEC—systems for legal metrology, weights and measures, and calibration. Once these "building blocks" are in place, APEC should move to conclude agreements on acceptance of test data and laboratory accreditation.

There has been discussion within APEC concerning the forum's approach to trade facilitation, with some member economies advocating simply the encouragement of bilateral MRAs between member states. While this is a laudable goal, APEC should concentrate its early, limited goodwill and resources on programs to engage *all* members in APEC-wide agreements. Unilateralism as a concept for APEC's standards work is not enough reason for APEC economies to sustain progress to 2020. Finally, for APEC to be successful, it needs a broader number of priority sectors for MRAs by 2020 than toys and food products, the two sectors currently under discussion. APEC needs to make progress in important infrastructure-related sectors, such as construction materials, transportation vehicles, and other capital goods. Reaching agreement in these sectors would serve to meet trade goals of both developed and developing members of APEC.

Other Priority Action Items under Discussion

In addition to the action items discussed above, APEC is consulting on how to promote MRAs in voluntary sectors. APEC should concentrate most of its limited resources on regulated sectors. It can, however, provide public endorsement to private sector-led agreements. To support industry, APEC should move quickly to leverage agreements already in force among private organizations in Asia. There are laboratories and certification organizations (for nonregulated products) operating in the region under bilateral MRAs. There are also MRAs in existence between accreditation bodies within APEC. These include agreements between TELARC in New Zealand and A2LA in the United States, TELARC and HOKLAS in Hong Kong, A2LA in the United States and HOKLAS, several agreements between New Zealand and Australia, and A2LA in the United States. There are also talks on MRAs between accrediting quality management systems and product certification organizations within the APEC region. The private sector Asia-Pacific Laboratory Accreditation Cooperation (APLAC) is discussing MRAs on laboratory accreditation within APEC. APEC should work creatively to engage private organizations, such as those above, in its work.

Finally, APEC is discussing methods to promote alignment of APEC member national standards with international standards. Priority areas under discussion include: (1) air conditioning, TV sets, and refrigerators, (2) food labeling, (3) rubber gloves and rubber condoms, and (4) unplasticized plastic pipes and fittings product. There may be merit in this type of work program for APEC, especially if it involves program to aid developing economies in standards development *before* national standards are developed. To the extent standards are written with reference to international standards, there are clearly benefits to trade. Much information on the trade implications of standards that are already in place, but not aligned with international ones is, however, lacking. APEC should concentrate on gathering data and information in this area and rely on private industry to decide, in most cases, where and when to harmonize to international standards.

CONCLUSION

The APEC agenda to 2020 on standards holds great promise as one of the most significant and beneficial parts of the organization's move toward open trade. By supporting implementation of major new elements in the Uruguay Round, aiding in the modernization of the region's technical infrastructure, and streamlining duplicate testing and certification requirements, APEC will help facilitate not only regional trade and economic development, but also support progress in world trade in the post–Uruguay Round environment.

NOTES

The conclusions and recommendations in this chapter are those of the author and do not necessarily represent those of the National Academy of Sciences, National Research Council, or other organizations.

1. For a discussion of these issues, see: *Standards, Conformity Assessment, and Trade: Into the 21st Century* (Washington, D.C.: National Research Council, National Academy Press, 1995).

2. For an overview of the SPS Agreement see: *Sanitary and Phytosanitary Safety Standards for Foods in the GATT Uruguay Round Accords*, Donna U. Vogt, Congressional Research Service (94-512-SPR), Library of Congress, June 21, 1994, Washington.

3. or an analysis of the 1994 Agreement see: Steinberg, Richard H. 1994. "The Uruguay Round: A Legal Analysis of the Final Act," *International Quarterly* 6(2):1–97.

4. Calculated using World Bank, IMF data on imports for 1991, totals for signatories to the Tokyo Round Standards Code, and new signatories to the Uruguay Round Agreement.

5. Data included in Report to Congress on Recommendations on Future Free Trade Area Negotiation, July 1, 1994, Table 2.

6. Based on estimates by NRC Project on International Trade, Standards, and U.S. Trade Policy project. Calculated from data supplied by the U.S. Bureau of the Census, 1993–1994.

7. For an overview of efforts to develop environmental management system standards at the international level see: Block, Marilyn R. 1994. "ISO/TC 207: Developing an International Environmental Management Standard," *The European Marketing Guide: Economic, Environmental, Legal & Social Strategies. Part of The Complete European Digest.* March 1994(II)3. (Atlanta, Georgia: SIMCOM).

8. For an overview of the ISO 9000 standard see: Breitenberg, Maureen. 1993. *More Questions and Answers on the ISO 9000 Standard Series and Related Issues.* NISTIR 5122. Prepared for the National Institute of Standards and Technology, U.S. Department of Commerce. Gaithersburg, Maryland: NIST, and ISO 9000 Central Secretariat. 1992. *ISO 9000 International Standards for Quality Management—Compendium.* 2nd ed. Geneva: International Organization for Standardization.

9. See also; Breitenberg, Maureen. 1988. *The ABC's of Certification Activities in the United States.* NBSIR 88-3821. Prepared for the National Bureau of Standards, U.S. Department of Commerce. Gaithersburg, Maryland: NBS.

10. For an overview of the evolution of the U.S. conformity assessment system framework, and its link to changes in manufacturing and business practices, see John W. Locke, *Conformity Assessment—At What Level?*, presentation to the Joint ISO, ANSI, and ASQC ISO 9000 Forum Application Symposium, Washington, D.C., October 7, 1993.

11. Based on data from the Bureau of the Census, U.S. Department of Commerce.

12. See also, Hyer, Charles W., ed., 1991, *Directory of State and Local Government Laboratory Accreditation/Designation Programs.* NIST Special Publication 815. U.S. Department of Commerce. Gaithersburg, Maryland: NIST. See also, Hyer, Charles W., ed., 1992, *Directory of Professional/Trade Organization Laboratory Accreditation/Designation Programs.* NIST Special Publication 831. U.S. Department of Commerce. Gaithersburg, Maryland: NIST.

13. Trade Policy Coordinating Committee, Product Standards Working Group, estimates applied to U.S. Census Bureau export data.

14. Data supplied by Charles Ludolph, director, Office of European Affairs, International Trade Administration, U.S. Department of Commerce, 1994.

15. For a discussion of the early results of these talks see; Ludolph, Charles M., 1994, "Mutual Recognition Agreements - Access to the European Union," *The European Report on Industry; Quality and Standards. Part of The Complete European Digest.* March 1994(II)3. (Atlanta, Georgia: SIMCOM).

16. For an overview of the possible benefits of such an MRA on medical devices see: *EU-U.S. Mutual Recognition Agreements (MRAs), Key Issues for the Medical Device Industry.* Washington, D.C. Health Industry. Manufacturers Association, 1994.

17. Estimated by R.B. Toth Associates, Washington, D.C.

18. Material provided by Walter G. Leight, 1994, Technology Service, National Institute of Standards and Technology, U.S. Department of Commerce.

19. See *A Vision for APEC, Towards an Asia Pacific Economic Community*, Report of the Eminent Persons Group to APEC Ministers, October 1993, pp. 44-45.

REFERENCES

Agreement on Technical Barriers to Trade. Uruguay Round, MTN/FA II-A1A-6, 1993.

Besen, Stanley M., and Leland L. Johnson. *Compatibility Standards, Competition, and Innovation in the Broadcasting Industry*. National Science Foundation (R-3453-NSF), 1986.

Block, Marilyn R. "ISO/TC 207: Developing an International Environmental Management Standard." *The European Marketing Guide: Economic, Environmental, Legal & Social Strategies. Part of The Complete European Digest*. Atlanta, Ga.: SIMCOM, March 1994.

Breitenberg, Maureen. *More Questions and Answers on the ISO 9000 Standard Series and Related Issues* (NISTIR 5122). National Institute of Standards and Technology, U.S. Department of Commerce. Gaithersburg, Md.: NIST, and ISO 9000 Central Secretariat, 1993.

———. *ISO 9000 International Standards for Quality Management—Compendium*. Geneva, Switzerland: International Organization for Standardization, 1992.

———. *Directory of Federal Government Laboratory Accreditation/Designation Programs* (NIST Special Publication 808). U.S. Department of Commerce, Gaithersburg, Md.: NIST, 1991a.

———. *Laboratory Accreditation in the United States* (NISTIR 4576). Gaithersburg, Md.: NIST, 1991b.

———. *The ABC's of Certification Activities in the United States* (NBSIR 88-3821). National Bureau of Standards, U.S. Department of Commerce. Gaithersburg, Md.: NBS, 1988.

Farrell, Joseph, and Garth Saloner. "Competition, Compatibility, and Standards: The Economics of Horses, Penguins and Lemmings." *Product Standardization and Competitive Strategy*, H. Landis Gabel. Amsterdam, Netherlands: Elsevier Science Publishers B.V., 1987.

Hankins, Ueanne, Jim Lockhart, Brett Snyder, and Jan Edwards. *Final Report of the Committee on National Accreditation of Environmental Laboratories*, Attachment #9, September 11, 1992.

Hyer, Charles W. *Directory of Professional/Trade Organization Laboratory Accreditation/Designation Programs* (NIST Special Publication 831). U.S. Department of Commerce, Gaithersburg, Md.: NIST, 1992.

———. *Directory of State and Local Government Laboratory Accreditation/Designation Programs* (NIST Special Publication 815). U.S. Department of Commerce, Gaithersburg, Md.: NIST, 1991.

Locke, John W. *Conformity Assessment—At What Level?* American Association for Laboratory Accreditation. Joint ISO, ANSI, and ASQC ISO 9000 Forum Application Symposium, Washington, D.C. Oct. 7, 1993.

Ludolph, Charles M. "Mutual Recognition Agreements—Access to the European Union." *The European Report on Industry: Quality and Standards. Part of The Complete European Digest*. Atlanta, Ga.: SIMCOM, March 1994.

National Research Council. *Standards, Conformity Assessment, and Trade: Into the 21st Century.* Washington, D.C.: National Academy Press, 1995.

Steinberg, Richard H. "The Uruguay Round: A Legal Analysis of the Final Act." *International Quarterly*, 1994.

Vogt, Donna U. *Sanitary and Phytosanitary Safety Standards for Foods in the GATT Uruguay Round Accords.* Congressional Research Service (94-512-SPR), Washington, D.C.: Library of Congress, June 21, 1994.

Wilson, John Sullivan. "Standards, Conformity Assessment, and Trade." *Asia Pacific Economic Cooperation: Theory and Policy.* JAI Press, 1995.

11

Japan's Double Standards: Technical Standards and U.S.–Japan Economic Relations

Brian Woodall

TECHNICAL STANDARDS AND TRADE FRICTION

A country's technical standards[1] sanitary restrictions, testing and certification, and approvals constitute political institutions with implications that transcend national boundaries. Domestically, technical standards specify the rules of the game concerning the testing, production, marketing, and distribution of products and services in home markets. As an Office of Technology Assessment (OTA) study neatly puts it, "Standards help determine the efficiency and effectiveness of the economy, the cost, quality, and availability of products and services, and the state of the nation's health, safety, and quality of life" (OTA, 1992, p. 7). Internationally, by defining the rules of the game in overseas markets, technical standards are able to create trade or divert it. In other words, technical standards can enhance or reduce welfare gains for the domestic economy (or regional bloc), the economies of trading partners, or the world economy. Trade disputes often arise when a country's technical standards are perceived to unfairly disadvantage or exclude foreign products or services. This chapter focuses on Japan's technical standards, the domestic context in which they are forged, and their implications for United States–Japan economic relations.

Before embarking on this analysis, it is necessary to define the phenomenon under scrutiny. A National Research Council study defines a "standard" as "a prescribed set of conditions or requirements concerning the definition of terms; specification of performance, operation, or construction; delineation of procedures; or measurement of quantity and/or quality in describing features of products, processes, systems, interfaces, or materials" (NRC, 1995, p. 205). But this definition is too broad for our purposes; therefore, it is essential to narrow the focus to embrace those facets of technical standards that relate to government policy and trade. A "mandatory standard" is a technical standard set by government, or a voluntary standard developed for private use, that is explicitly or implicitly recognized in government procurement or regulatory policies (NRC, 1995, p. 205). Since the vast majority of mandatory standards do not generate trade disputes, however, it is necessary to further fine-tune the definition. A "discriminatory standard," the focus of this analysis, is a standard set, or recognized by a government, that excludes or unfairly disadvantages the products or services of foreign firms. In other words, a discriminatory standard is a standard that is set differently from widely accepted international standards and specified in such a way as to exclude or unfairly handicap foreign products or services.

This chapter draws upon accounts of a series of complaints about Japanese technical standards that have been featured in U.S.–Japan trade friction. Importantly, the bulk of these American allegations involve Japanese standards pertaining to distressed industries, especially those represented by powerful domestic lobbies, and industries deemed strategic from the standpoint of the national interest. Thus, a great portion of this American ire has concentrated on technical standards in Japanese industries and sectors at opposite ends of the product life cycle. As argued in this chapter, a focal point in the bilateral trade conflict has been Japan's system of double standards.

The analysis that follows spotlights the motives and behavior of the policy protagonists in the standards-setting process and traces the roots of Japan's double standards to the country's dual-structure political economy. On this score, it is important to note that discriminatory standards are not the exclusive preserve of the Japanese; indeed, the United States and the European Community employ standards and testing requirements that could be classified as nontariff barriers.[2] It bears mentioning that Japan is a signatory to the Agreement on Technical Barriers to Trade[3] and boasts membership in the International Organization for Standardization and the International Electrotechnical Commission, major international standards-setting bodies. It is hoped that this worm's-eye view of Japanese standards and standards-setting will inspire further comparative analysis. Moreover, one need not accept at face value all of the American allegations concerning Japan's standards. Nevertheless, the list of allegations is sufficiently lengthy and the circumstantial evidence sufficiently strong to justify this country-specific analysis. The objective in this chapter is to identify characteristic patterns and to draw some admittedly tentative conclusions concerning the nature, sources, and implications of Japanese technical standards for bilateral trade relations.

JAPAN'S DISCRIMINATORY STANDARDS: A SAMPLING

Even a cursory survey of the recent secondary literature on U.S.–Japan economic relations provides an extensive list of allegedly discriminatory technical standards in Japan (inter alia, Prestowitz, 1989; Lincoln, 1990; Bergstrom and Noland, 1993; and Bayard and Elliott, 1994). Before examining some of these cases, it is well to call attention to the broad range of different industries and sectors in which discriminatory standards have appeared. Indeed, citings of such have been made in high-tech as well as low-tech industries, and in "sunrise" as well as "sunset" sectors. Moreover, virtually every ministry and agency of the Japanese government has been implicated in some sort of bilateral trade dispute involving technical standards. Clearly, therefore, the most likely culprit, the Ministry of International Trade and Industry—so-called notorious MITI, the chief executive officer of "Japan, Inc."—is not the only suspect.

Trade disputes concerning Japanese standards have arisen in a host of different venues. Perhaps the classic illustration, at least in the mind of the American public, depicts a Japanese customs inspector standing at the dock, chalk and ruler in hand, rejecting the entry of U.S.-made automobiles because of minuscule and merely cosmetic unevenness in side mountings. Trade disputes have been sparked by Japanese standards that barred the importation of American-made aluminum baseball bats and foreign ski equipment. Japanese "uniqueness"—i.e., the allegedly unique brand of baseball played in Japan and the supposedly *sui generis* nature of the snow flakes that fall on the country's ski slopes—was the ostensible reason for keeping out these products. Similarly, foreign construction firms—including international giants like Bechtel and Fluor Daniel—were initially prohibited from submitting bids on projects to construct the $8 billion New Kansai International Airport because of the allegedly unique characteristics of the mud on the floor of Osaka Bay. In addition, a long-standing, but now discontinued, customs' practice mandated that all imported flower bulbs be sliced in half to inspect for parasites. As Bagwati observed, "even Japanese ingenuity could not put them back together" (Bergstrom and Noland, 1993, p. 72).

Safety and testing standards feature prominently in the laundry list of foreign complaints. For example, Japanese standards for processed foods, formaldehyde levels in infant clothing, and the durability of small fiberglass boats have come under fire. In addition, a former Commerce Department official recounts the obstacles encountered by a major American cosmetics manufacturer in attempting to market a hair dye in Japan. After U.S. market opening pressure secured publication of a list of approved ingredients (which had hitherto been kept secret), the firm began producing the new product using only stipulated ingredients and sent samples to the Ministry of Health and Welfare (MHW) for approval. Yet even that step would have been unnecessary for marketing a product in the United States, where the Food and Drug Administration requires only package listing of approved ingredients and does not interfere unless some complaint arises. In the end, it required many months of costly delay, an MHW request for reformulation and resubmission, and, ultimately, pressure from the U.S. government in order to obtain the ministry's approval of a hair dye product containing only sanctioned ingredients (Prestowitz, 1989, pp. 211–212).

Safety standards have been the focal point in a series of bilateral confrontations over forestry and paper products. The thrust of the American government efforts have aimed at pressuring the Japanese government to amend the country's restrictive building codes, permit the participation of foreign firms in the standards-setting process, and designate foreign testing organizations to provide certification on exports to Japan (Lincoln, 1990; Bayard and Elliott, 1994). Contentions focused on restrictions on the use of two-by-four lumber (the standard used in U.S. construction), bans on imports of ponderosa and lodgepole pine products, and size restrictions on family dwellings. Backed by pressure from the National Forest Products Association and the industry's champions in congress, the United States Trade Representative (USTR) ultimately initiated an investigation under "Super 301" of the Trade Act of 1988. More mention is made later concerning the case of Japanese standards for forestry and paper products.

Complaints have also arisen over Japan's phlegmatic approval process and allegedly unnecessary restrictions on pharmaceutical products and medical equipment. In some instances, the cause of the dispute has been MHW's official medical fee schedule, a list containing set fees for the many thousands of drugs and medical procedures for which medical providers are able to receive reimbursement. Under Japan's universal, pay-for-service medical system, providers are not reimbursed for a product or service unless it is listed on the fee schedule. Not surprisingly, the process of securing listing of a product on the fee schedule and gaining approval is both lengthy and, on occasion, murky. The U.S. government has pressured the Japanese government to accept foreign clinical test data and streamline the approval of reagents for drugs, and to clarify areas in which approval is or is not required for medical equipment (Lincoln, 1990, p. 150).

In the case of heat-treated blood products, the phlegmatic approval process may have produced tragic results. Investigative reporting by the *Mainichi Shinbun*, a major Japanese daily, found that the Health Ministry finally approved the commercialization of heat-treated blood products in July 1985, more than two years after such products had become available in America. Whether by coincidence or not, a major Japanese firm began marketing heated blood products only one month following this belated approval, and unheated blood products continued to be imported and remained on hospital shelves until 1986. Whatever the reasons for the delay, this resulted in the spread of the AIDS virus among Japanese hemophiliacs, with all of the predictable and tragic consequences (Mainichi Shinbun, 1992).

The procurement standards of the Japanese government have also come under fire. In a heated bilateral trade confrontation, American producers alleged that the Nippon Telephone and Telegraph (NTT), then a quasi-governmental organ with monopoly control over the domestic telecommunications market, employed procurement criteria designed to exclude foreign producers. One of the more memorable barbs hurled in that particular brouhaha was uttered by the president of NTT, who insisted that if Japan's telecommunications giant purchased anything at all from American firms it would be "mops and buckets" (Curran, 1982, p. 201). More recently, bilateral relations became soured over allegedly discriminatory standards in Japan's cellular phone market. Government procurement standards also came under fire in an ongoing bilateral rift over access to Japan's enormous construction market. In that instance, American firms found themselves unable to submit bids for Japanese public works projects because of a stipulation that required that all would-be bidders undergo assessment based on the results of construction works during the previous two years within the Japanese market. Since no American construction firm had performed work in the Japanese market in nearly two decades, this devious catch-22 embodied a de facto market access barrier (Woodall, 1996).

TECHNICAL STANDARDS AND U.S.–JAPAN TRADE FRICTION

Allegations of discriminatory technical standards have become a familiar refrain in U.S.–Japan trade conflict. As Lincoln explains:

Standards have been shaped in cooperation with domestic interests specifically to exclude foreign products; testing agencies have been controlled by domestic manufacturers or have refused to visit foreign factories; patent approvals have been delayed deliberately to allow domestic firms to create a competing product; government-funded agencies have been told bluntly that purchase of foreign items would jeopardize their funding; and trading companies have been warned to moderate their purchases of certain foreign products. (1990, pp. 14–16)

On this score, the American government's obsessive quest to dismantle unfair standards may be seen as part of a larger effort to pry open closed Japanese markets and "level the playing field" so as to redress the nagging bilateral trade imbalance. This market-access offensive gained momentum in the mid-1980s with the rapid appreciation of the yen following the 1985 Plaza Accord and the phenomenal expansion of asset values in the midst of Japan's "bubble economy."

Bilateral disputes over Japanese technical standards date back to 1970s. The Tokyo Round of the General Agreement on Tariffs and Trade (1974–1979) brought pressure to bear on Japan to eliminate nontariff barriers, including discriminatory technical standards. In the midst of this, the issue of unfair standards in forestry and paper products was taken up in talks between United States Trade Representative Robert Strauss and Foreign Minister Nobuhiko Ushiba. The 1978 Strauss-Ushiba communiqué called for the removal of barriers to increased bilateral trade in forestry products (Bayard and Elliott, 1994). Shortly thereafter, Japan became a signatory of the Standards Code administered under the General Agreement on Tariffs and Trade.

Standards again became an agenda item in the Market-Oriented Sector Selective (MOSS) talks that began in the mid-1980s. Sectoral access problems involving Japan's standards featured prominently in each of the four targets of the American assault: telecommunications equipment and services, medical equipment and pharmaceuticals, wood and paper products, and software. The rationale of the U.S. side in targeting these specific sectors was the perception that, in each sector, American firms possessed superior technology and competitive advantages vis-à-vis their Japanese competitors. In the absence of many significant formal barriers, therefore, nontariff barriers such as discriminatory standards must be the culprit in denying access to American firms in these disputed Japanese markets. Or so assumed the American side in pressing its case in the MOSS talks.

Technical standards appeared on the American negotiating team's "hit list" in the Structural Impediments Initiative (SII) talks that began in the late 1980s. The main agenda items included land prices that made it prohibitively expensive for foreign firms to do business in Japan, impediments to foreign direct investment emanating from the *keiretsu* system of interlocking directorates and cross-shareholding, a byzantine distribution system, the retarded state of the country's road and sewerage systems and other infrastructure, and a limp-wristed antitrust law. One prong in the American offensive in the SII talks was to dismantle access barriers, such as the aforementioned catch-22 element in the designated bidder system, in Japan's public construction market. In this case, United States Trade Representative Clayton Yeutter and Special Envoy Ichiro Ozawa produced the 1988 Major Projects Agreement that altered bidder designation qualifications for foreign firms on a few specified large-scale public works projects (Woodall, 1996).

Finally, allegations of discriminatory technical standards surfaced yet again in the "Framework Talks" initiated by the Clinton administration. Those talks spotlighted allegedly unfair Japanese standards for telecommunications equipment, automobiles and auto parts, medical technology, and insurance services. As late as July 1993 the position of the U.S. government held that "The Government of Japan maintains highly detailed standards for the manufacture, performance, and labeling of products. Numerous cases exist in which Japanese standards were written or changed to disadvantage or exclude imported products, often after imports had begun significantly to penetrate the market" (National Trade Estimates Report, 1993). Even though the Japanese government has simplified, harmonized and, in some cases, eliminated discriminatory standards, the Clinton administration's "results-oriented" approach to bilateral trade assures that disputes over standards and other nontariff barriers will not soon disappear from the negotiating agenda.

JAPAN'S STANDARDS AND STANDARDS-SETTING

Japan employs a myriad of technical standards, specifications, and market regulations. One source lists sixty-three separate standards-related laws, grouped into ten different categories: protection systems (4 laws); food, liquor, tobacco, and pharmaceuticals (7 laws); household products (4 laws); industrial products 1 law); dangerous objects and safety regulations (16 laws); services (15 laws); intellectual property (4 laws); procedures for export and import (6 laws); public procurements (2 laws); and miscellaneous (4 laws) (*Guide*, 1993). These laws dictate standards and specify restrictions and specifications for all manner of things, ranging from food control and pharmaceuticals to antitrust and the services of foreign lawyers, and from government procurement and the labeling of household appliances to the control of firearms and swords. In addition, Japanese Industrial Standards (JIS) are voluntary national standards for industrial and mineral products. As of March 31, 1993, there were more than eight thousand four hundred such standards on the books, and some sixteen thousand permissions (or approvals in the case of foreign producers) had been given to factories allowing them to affix the JIS mark[4] on their products (*JIS Yearbook*, 1993).

In most cases, technical standards are determined by private-sector industrial associations and technical societies that interact with the responsible government ministries or agencies. Standards for products bearing the JIS mark are developed by the private sector and those government ministries under the aegis of the Ministry of International Trade and Industry (MITI) and the Japanese Industrial Standards Committee (JISC). In a typical case, government entrusts the task of preparing a JIS draft to an industrial association or technical society, which submits the finished draft to MITI's Agency of Industrial Science and Technology (AIST). On occasion, private think tanks or institutes affiliated with AIST are commissioned to conduct background research to assist in drafting the proposal. After drafting is completed, the proposal is sent to the JISC for deliberation and, if necessary, modification. Deliberation is carried out in one of JISC's eighteen divisions,[5] which are further subdivided into technical committees (*JIS Yearbook*, 1993, p. 8). If consensus is reached among the major concerned interests, the AIST announces the new JIS in the Official Gazette, and its text will be published by the Japan Standards Association.

Many actors play a role in the standards-setting process. Legal specialists, labor leaders, representatives of consumer groups, technical specialists and academics, and legislators contribute input. But the real protagonists in standards-setting are industrialists and government bureaucrats. Japan's private sector is subdivided into a plethora of well-organized industry associations. Most of these associations claim membership in some sort of sectoral umbrella organization that is subsumed under the Federation of Economic Organizations (Keidanren). In the case of construction, for example, the Japan Civil Engineering Contractors' Association (Nihon Doboku Kogyo Kyokai or Dokokyo), representing the interests of the largest general contractors, is but one of more than one hundred registered industry associations registered with the Ministry of Construction. Dokokyo, along with the similar associations representing railway and electrical power contractors, composes the backbone of the Japan Federation of Construction Contractors, which is one of Keidanren's member organizations (Woodall, 1990). These private-sector representatives fully understand the power of standards in shaping the competitive position of their industries and firms. Hence, it is logical to assume that strategic concerns— i.e., predictability in market competition and competitive advantage in domestic and-or international trade—constitute the lowest common denominator in driving the standards-setting decisions of these industrialists.

Government bureaucrats also play a significant role in the standards-setting process. For obvious reasons, a major role is played by AIST officials at MITI, but virtually every government ministry and agency has a stake in determining standards, specifications, and regulations. On this score, it is fair to assume that Japanese bureaucrats, like government officials elsewhere, are human beings who are at least not entirely motivated by the general welfare or the interests of the state (Niskanen, 1971). It is claimed that the Japanese were the first to recognize the potential of standards as key components of industrial policy. "Because Japan had a small domestic market, and was late in the process of industrialization, the Japanese Standards System (JSS) originally focused on improving economic efficiency and gaining the benefits of technology transfer. Later, standards were used to control product quality, and thereby promote trade" (OTA, 1992, p. 84). On this score, it is well to note the relatively "strong" and traditionally wide-ranging interventionary role played by Japanese state bureaucrats in all manner of affairs relating to the economy (Lockwood, 1968; Johnson, 1982; and Katzenstein, 1985).

Control over standards bestows power and enhances the prospects of pecuniary security for Japanese government officials. Bureaucrats use their extensive powers over licensing and approval, as well as technical standards, to guide the activities of the industries and firms in their respective bailiwicks[6] (Omiya, 1994, p. 112). In addition to these formal power, Japanese bureaucrats frequently use informal directives known as "administrative guidance" to apprise industries and specific firms of their intentions (OTA, 1992, p. 34). Often these firms offer "second careers" to ex-officials, the majority of whom choose, or are obliged, to retire from the government service between the ages of forty-five and fifty-five. This interpenetration of state officialdom and the private sector need not be interpreted as evidence of government's "capture" by the private sector; indeed, "descent from heaven" (*amakudari*), as the practice is known, is mutually beneficial (Woodall, 1996). In this regard, an important and largely unexplored topic concerns the presence and role of ex-bureaucrats in the advisory councils (*shingikai*) and think tanks that offer input on standards-setting decisions.

The point here is to spotlight the vested interests involved in the Japanese standards-setting process. Standards bestow pecuniary benefits and offer career security for industrialists and government bureaucrats. In so doing, standards and standards-setting constitute an important pillar in Japan's vaunted government-business relations and the interpenetration of the country's economic and administrative elite. Often the standards shaped by these vested interests come to be seen as nontariff barriers. Even in cases in which these barriers are not the product of conscious decisions to exclude foreign products or services, the effect is the same. The result is allegations of unfair trading practices and international conflict.

EXPLAINING JAPAN'S TECHNICAL STANDARDS

The accepted wisdom contends that regulatory barriers in standards-setting tend to emerge in one or the other of two settings.[7] First, it is held that preferential standards are more likely to be found in cartelized (or concentrated) industries, where it is relatively easy to identify and accommodate all interested parties, and where returns on investment are more likely (OTA, 1992, p. 10). And, second, it is pointed out that established primary products sectors with well-entrenched lobbies tend to wield the political clout necessary to extract tailor-made standards that reward favored interests and inhibit the entry of newcomers (Bergsten and Noland, 1993, p. 72). In sum, the prevailing thinking holds that discriminatory standards tend to be erected to protect the interests of cartelized industries or industries represented by politically potent lobbies. Little attention is given to the role of the state or conditions at the level of the world political economy in facilitating the rise and sustainability of discriminatory standards. As we shall see, while the standard wisdom explains part of the story of Japan's double standards, important questions remain unanswered.

Protective Standards

Numerous complaints have alleged that, either by design or by default, Japanese standards serve to protect the interests of cartelized industries or well-entrenched primary producer lobbies. Although these "protective standards" are often justified on economic grounds (i.e., employment for declining industries), as often as not their true rationale is political. A well-documented case of such regulatory protection involves the agricultural sector, with its nationwide organizational embrace and institutionalized ties to Japan's legislative elite (Anderson and Hayami, 1986; Hayami, 1990; and George, 1991). At various points in time, standards also have been employed to protect the textile, securities and banking, pharmaceutical, and construction industries, as well as the country's myriad of small retailers (Destler et al., 1979; Horne, 1985; Yoshikawa, 1989; Woodall, 1996; and Upham, 1993). In the case of pharmaceuticals, for example, foreign firms were prohibited by regulatory policy from applying on their own for the first step of drug approval, and clinical tests had to be conducted in Japan on native citizens (Reich, 1990).

A classic illustration of standards employed to protect a declining, but politically powerful, interest group is seen in the case of Japan's wood products and paper industry. Indeed, as noted earlier, one of the earliest and most enduring battlegrounds in U.S.–Japan conflict over technical standards involves forestry products. It has been argued that Japanese wood fabricators are high cost and inefficient by world standards, and comparative indications show that mill costs in Japan are double or triple those in the United States (Bayard and Elliott, 1994, p. 135). Despite the fact that Japan is the world's largest net importer of wood products, those imports have tended to be heavily skewed toward raw materials. The various barriers erected to protect domestic forestry firms have included discriminatory product standards; government tolerance of price-fixing; low-cost loans and tax breaks for inefficient producers; standards based on appearance rather than performance criteria; and excessively restrictive building and fire codes (Bayard and Elliott, 1994, p. 137). Specific examples include the lengthy battle to revise the building code to permit the use of two-by-four lumber products and the intriguing case of a foreign-made wood product coating with a clear fire retardant. When heated to a particular temperature, the retardant would turn to foam and douse a flame. However, Japanese law required that a product be subjected to a flame for a prescribed period before it could be certified as fireproof. Since the product in question extinguished the flame before the time expired, it was not given certification (Bayard and Elliott, 1994, pp. 141–142).

Japan's forestry industry wields political clout far in excess of its size or economic importance. Indeed, the number of Japanese forestry workers declined nearly 60 percent between 1965 and 1990, and, during roughly the same period, Japan's rate of self-sufficiency rate dropped from 90 percent to a mere one-quarter. One reason for the existence and persistence of these discriminatory standards is owed to a network of well-organized forestry cooperatives and to a badly malapportioned electoral system. Under the old electoral system (reformed in 1994), which elected from two to six representatives per district, candidates could be elected with less than 10 percent of the popular vote. This created an incentive for candidates to barter legislative influence in exchange for the pledge of support from an organized group purporting to deliver a bloc of votes. Thus, the political clout wielded by the wood and paper products industry, a player in the powerful agricultural lobby, was rewarded with protective standards that helped keep the products of foreign competitors out of Japanese markets.

Promotional Standards

As noted, the accepted wisdom maintains that discriminatory standards are created to protect the interests of powerful domestic lobbies or cartelized industries. In essence, the prevailing thinking holds that these influential private-sector interests dictate the desired regulatory policies, to a "captive" state. But, as the Japanese case amply demonstrates, discriminatory standards can also arise in a setting in which the state is sufficiently strong to set and enforce standards that aim to develop industries or sectors deemed strategic in the national interest.[8] In other words, "promotional standards"—standards that, by design or default, promote the growth of emerging industries and sectors—represent a qualitatively different category of technical standards[9] (Woodall, 1996). Japan's promotional standards and related policies are concerned with the development of strategic, high-growth industries and sectors, and the protection of infant industries deemed to have the potential of achieving economies of scale and significant technological spill-over effects.

It is important to recognize that, typical of a late-industrializing country, the Japanese state assumed extensive powers in developing the country's economy (Gerschenkron, 1962; and Dower, 1975). In this effort, the prewar state employed a range of policies to encourage the growth of particular industries and sectors such as textiles, steel, and shipbuilding. Technical standards were employed as a tool of "industrial rationalization policy" (Johnson, 1982). As Lecraw observes:

At the start of Japan's industrialization in the late 1800s, its industrial firms were small and inefficient, and lacked a modern technology base. To meet these problems, the Japanese government actively promoted industry rationalization, simplification of product variety, and interchangeability and compatibility between products. On the one hand, this strategy enabled Japanese firms to achieve efficiency of high volumes even though they were relatively small, and on the other, it facilitated the transfer of technology from abroad since the same product or process could be used by all firms within an industry and by firms across industrial sectors. (1987, p. 31)

But the use of standards as a tool for industrial rationalization and development has continued into the postwar era. On this score, standards have been featured in the state's developmental strategy, among them the machine tool, semiconductor, and telecommunication industries. Recently, a proposed software quality standard—the JIS Z9901 certificate[10]—generated apprehension among American business and trade officials fearful of a new barrier blocking access to Japanese markets (Choy, 1995, pp. 4–5).

The case of advanced medical technology illustrates some of the ways in which standards have been used to promote the growth of a strategic industry. Although, as of the mid-1990s, American firms were doing rather well in Japan's medical equipment market, relative to those in automobiles and auto parts, insurance, and government telecommunications procurement, complaints arose concerning discriminatory standards. For example, the U.S. producer of an implantable pain-killing device encountered what may have been unintentional discrimination when the Health Ministry refused in 1982 to list the product on the official fee schedule for reimbursement under the national health insurance system. It took ten years before approval was granted and sales could begin (Pollack, 1994). However, premeditation may well have figured in government standards-setting for pacemakers and brain scanners, markets in which Japanese firms lag far behind foreign competitors. In the case of pacemakers, price controls were imposed at about the same time a series of highly publicized arrests were made in which Japanese dealers were accused of bribing physicians to win sales. Similarly, the American maker of an advanced brain scanner, the only model approved for medical use in Japan at the time, failed in two bids to gain procurement by a MITI-affiliated research institute. Whether by coincidence or not, the fact that MITI was backing a consortium aimed at developing a similar technology added fodder to allegations of discriminatory standards (Pollack, 1994).

DOUBLE STANDARDS AND A DUAL POLITICAL ECONOMY

In sum, two types of technical standards—protective as well as promotional—are employed as weapons in Japan's industrial policy arsenal. The reason why these double standards take the form and serve the functions they do is rooted in the very structure of Japan's "dual political economy." Contrary to popular perceptions about a monolithic "Japan, Inc.," therefore, the case of technical standards reveals a fundamental duality in Japanese policy objectives, processes, and outcomes.

The saying "first-rate economy, third-rate politics" refers to the paradoxical coexistence of world-class levels of economic performance in Japan's internationally competitive sectors, in contrast to Third World levels of inefficiency and political clientelism in certain politically vital domestic domains (Woodall, 1996). Much has been written concerning Japan's efficient and relatively unpoliticized internationally traded sectors and industries. Recently, however, market access problems have attracted attention to the inefficient, highly protected, and patronage-ridden distributive sectors of Japan's political economy. Notable among these are the government-set rice price, the protection for small business, the official tolerance of a multitiered distribution system, a "nonpolicy" for land use, and a rigged public construction market. "After reviewing Japan's domestic policy sectors," Calder observes, "one begins to wonder both where the economic juggernaut so clearly visible to foreign competition could have come from and why domestic policy patterns so differ from the common generalizations about Japanese policymaking in internationally traded sectors" (1988, p. 465).

Economic policy in Japan is forged in two fundamentally different policy markets. A "policy market" is the stage on which the key political actors meet and make key policy decisions. Out of the process of political exchange emerge the public policies that determine the allocation of economic and political benefits for the society. Policies for potential growth sectors, internationally competitive industries, and sectors deemed vital to the national interest emerge from a "strategic" policy market. Government bureaucrats and private sector-elites act as the key policy protagonists, while elected politicians play a decidedly minor part. Meanwhile, the largest share of policymaking takes place in a highly politicized "structural" policy market encompassing domestic sectors in which the public is often the client. In these domains, elected politicians as well as government bureaucrats and private-sector elites play a direct and extensive role in policy formulation.

In this sense, the Japanese system is a "dual political economy" in which segmented policy markets generate separate but interactive "policy regimes" that operate under different equilibrium conditions. These segmented policy regimes may be referred to as the "developmental state" and the "clientelist state." Promotional standards are among the various policy instruments—including tariff, tax, and financial incentives—used to promote the growth of industries deemed strategic by the developmental state. At the same time, the clientelist state employs protective standards to enhance the well-being of politically powerful lobbies or cartelized domestic interests and to shield those interests from the rigors of real market competition. Concerns of economic rationality dominate in the former, while political rationality tends to dictate outcomes in the case of the latter. And an unmistakable streak of economic nationalism is visible in both protective as well as promotional standards.

SUMMARY AND IMPLICATIONS

Technical standards are critical political institutions that, among other things, spell out the rules of the game for domestic firms, as well as the terms of competition for foreign companies seeking market penetration. This study has sought to illuminate the nature and ramifications of the phenomenon of the discriminatory standard, a mandatory standard set by or recognized by government that excludes or unfairly disadvantages foreign products or services. Empirical insights were drawn from an intensive analysis of a single country case, that of Japan. It was found that Japan employs a system of double standards: protective standards aimed at preserving the position of powerful domestic lobbies and promotional standards that seek to develop strategic industries and sectors. These double standards are forged in a dual political economy whose effective functioning depends upon the symbiotic interaction of two separate policy regimes. Foreign trade negotiators and business interests must appreciate this structural duality in formulating effective strategies to gain entree to Japanese markets.

In conclusion, it is appropriate to consider how Japan has sustained this system of double standards and how much longer it can continue to do so. As long as foreign firms, particularly American companies, displayed little interest in penetrating Japanese markets, it was relatively easy to employ standards that rigged domestic markets to favor selected firms and prohibited foreign competitors from snatching a significant market share in strategic industries and sectors. In large measure, therefore, the sustainability of this system of double standards depends upon the ignorance, apathy, or weakness on the part of disadvantaged domestic interests (i.e., unannointed firms or exploited consumers) or trade monitors in foreign countries, especially the United States. However, in an increasingly interdependent world economy, characterized by aggressive unilateralism and rising consumer activism, it is becoming more and more difficult to disguise regulatory barriers such as discriminatory standards. These changed realities of a new world order, coupled with a chronic and enormous international trade surplus, guarantee that Japan's double standards will continue to draw intense fire in trade disputes.

NOTES

I gratefully acknowledge the able research assistance supplied by Jason Godwin, who contributed to the completion of this study.

1. For the sake of brevity, I use the term "technical Standards" throughout this chapter to denote technical standards, sanitary restrictions, testing and certification, and approvals.

2. For example, a case might be made that Food and Drug Administration testing and certification procedures, or QS 9000, a standard developed by U.S. auto makers and applied to automobile parts suppliers, constitute discriminatory standards. Examples of potential discriminatory standards in the EC are discussed in OTA (1992, pp. 61-74); and Egan (1994).

3. Japan ratified the Agreement on Technical Barriers to Trade in April 1980. The Standards Code, as the agreement is popularly known, is designed to eliminate the use of standards, technical regulations, and certifications as nontariff barriers to trade.

4. JIS are divided into three categories: product standards (4,000 items), working method standards (1,600 items), and basic standards (2,800 items).

5. The 18 JIS divisions (and the number of JIS in force as of 31 March 1993) include civil engineering and architecture (508); mechanical engineering (1,308); electronic and electrical engineering (794); automotive engineering (341); railway engineering (215); shipbuilding (540); ferrous materials and metallurgy (327); nonferrous

metals and metallurgy (399); chemical engineering (1,593); textile engineering (308); mining (202); pulp and paper (93); ceramics (246); domestic wares (236); medical equipment and safety appliances (304); aircraft and aviation (91); information processing (216); and a grab-bag division containing packaging, welding, and radioactivity (685).

6. In 1994, the number of licensing and approval powers of eleven of the twelve government ministries were as follows: Transport (1966), International Trade and Industry (1915), Agriculture (1357), Finance (1236), Health and Welfare (1170), Construction (870), Labor (579), Education (322), Posts and Telecommunications (313), Home Affairs (114), and Foreign Affairs (50).

7. For purposes of this analysis, it is unnecessary to present a detailed analysis of the role of technical standards in enhancing economic efficiency, facilitating (or at least not stifling) innovation, contributing to product safety and public health, and guarding the environment. Nor is it necessary to trace the forces that generate a popular demand for or acceptance of government involvement in standards-setting, typically a pivotal event (e.g., America's wartime production and national campaign to reduce waste) or a health risk caused by an unsafe product (e.g., the Tylenol tampering panic). In this regard, standards are "impure" public goods. That is, they combine aspects of public goods (e.g., national defense) that are available to everyone and from which no one can be excluded, while at the same time serving functions for particularistic private-sector interests. It is sufficient for our purposes to note that the costs and difficulties of monitoring create an incentive for "free riding" and other collective action problems that tend to result in the underproduction of standards (Olson, 1971; and OTA, 1992, p. 9).

8. While Japan *may* have pioneered the practice (and, in so doing, dictated emulation elsewhere), it is not alone in employing promotional standards. European countries have long used standards as a tool of industrial policy (OTA, 1992). In the case of America, strategic concerns could well have figured in the Federal Communications Commission's regulation that high-definition television transmission in the United States either be receiver compatible or allow simulcast with existing broadcast signals. That 1988 regulation improved the competitive position of American firms at the expense of the Japanese research team, which had already developed a working—but incompatible—prototype. In addition, the National Competitiveness Act of 1993 recommended the use of standards to technological development in American industry. Jensen and Thursby (1994) offer an insightful study of the impact of standards set for strategic reasons prior to successful product development.

9. Domestic concerns determine which industries and sectors are deemed "vital" from the perspective of a given country's economic and security interests. For example, defense industries are deemed vital to the national interests of the United States, whereas the development of commercially viable industries is the principal concern in Japan.

10. Under the proposal, criteria for qualifying for a JIS Z9901 certificate depend upon whether or not: 1) production and service responsibilities are clearly defined; 2) the software is subject to preapproval or inspection techniques; 3) the software is documented as it is developed; and 4) the relevant software manuals use clearly defined terms.

REFERENCES

Anderson, Kym, and Yujiro Hayami. *Agricultural Protection: East Asia in Comparative Perspective*. London, England: Allen and Unwin, 1986.

Bayard, Thomas O., and Kimberly Ann Elliott. *Reciprocity and Retaliation in U.S. Trade Policy*. Washington, D.C.: Institute for International Economics, 1994.

Bergstrom, C. Fred, and Marcus Noland. *Reconcilable Differences? United States-Japan Economic Conflict*. Washington, D.C.: Institute for International Economics, 1993.

Calder, Kent E. *Crisis and Compensation: Public Policy and Political Stability in Japan, 1949-1986*. Princeton, N.J.: Princeton University Press, 1988.

Choy, Jon. "Americans Wary of Proposed Japanese Software Quality Standard." *Japan Economic Institute Report*, 24 B: pp. 4-6, 1995.

Curran, Timothy J. "Politics and High Technology: The NTT Case." *Coping with U.S.-Japan Economic Conflicts* (I. Destler and Hideo Sato, eds.). Lexington, Mass.: D.C. Heath, 1982.

Destler, I., Haruhiro Fukui, and Hideo Sato. *The Textile Wrangle: Conflict in Japanese-American Relations, 1969-1971*. Ithaca, N.Y.: Cornell University Press, 1979.

Dower, John W. *Origins of the Modern Japanese State: Selected Writings of E. H. Norman*. New York: Random House, 1975.

Egan, Michelle. "The Politics of European Regulation: Bringing the Firm Back In." Paper presented at the International Conference for Europeanists, Council for European Studies, 1994.

George, Aurelia. "The Politics of Interest Representation in the Japanese Diet: The Case of Agriculture." *Pacific Affairs*, 64: pp. 506-528, 1991.

Gerschenkron, Alexander. *Economic Backwardness in Historical Perspective*. Cambridge, Mass.: Harvard University Press, 1962.

Guide to Obtaining Information on Standards, Certification, and Other Regulations in Japan. Mimeo on file with the author, 1993.

Hayami, Yujiro. *Japanese Agriculture Under Siege: The Political Economy of Agricultural Policies*. London, England: Macmillan, 1990.

Horne, James. *Japanese Financial Markets*. London, England: Allen and Unwin, 1985.

Jensen, Richard, and Thursby, Marie. "Patent Races, Product Standards, and International Competition." Manuscript on file with the author, 1994.

JIS Yearbook. Tokyo, Japan: Japanese Industrial Standards Committee, 1993.

Johnson, Chalmers. *MITI and the Japanese Miracle: The Growth of Industrial Policy, 1925-1975*. Stanford, Calif.: Stanford University Press, 1982.

Katzenstein, Peter J. *Small States in World Markets: Industrial Policy in Europe*. Ithaca, N.Y.: Cornell University Press, 1985.

Lecraw, Donald J. "Japanese Standards: A Barrier to Trade?" *Product Standardization and Competitive Strategy* (H. Landis Gabel). New York: Elsevier Science Publishers, 1987.

Lincoln, Edward J. *Japan's Unequal Trade*. Washington, D.C.: Brookings Institution, 1990.

Lockwood, William W. *The Economic Development of Japan: Growth and Structural Change*. Princeton, N.J.: Princeton University Press, 1968.
Mainichi Shinbun. *Kakusareta AIDS: Sonotoki Seiyaku Gaisha, Koseisho, Ishi wa Nani o Shitanoka* (AIDS Cover-Up: The Role of Foreign Drug Companies, the Health Ministry, and Physicians). Tokyo, Japan: Mainichi Shinbun Sha, 1992.
National Research Council. *Standards, Conformity Assessment, and Trade: Into the 21st Century*. Washington, D.C.: National Academy Press, 1995.
Niskanen, William A. *Bureaucracy and Representative Government*. Chicago: Aldine Atherton, 1971.
Olson, Mancur. *The Logic of Collective Action: Public Goods and the Theory of Groups*. Cambridge, Mass.: Harvard University Press, 1971.
Omiya, Kenichiro, and Group B. *Ji-sha Renritsu Seiken Seijika-Kanryo Jinmyaku Chizu* (A Map of the Networks of Japan's Politicians and Bureaucrats). Tokyo, Japan: Yubasha, 1994.
Pollack, Andrew. "Medical Companies Berate Japan." *New York Times*. March *Regulations in Japan*. Mimeo on file with the author, 1994.
Prestowitz, Clyde V., Jr. *Trading Places: How We Are Giving Our Future to Japan and How to Reclaim It*. New York: Basic Books, 1989.
Reich, Michael R. "Why the Japanese Don't Export More Pharmaceuticals: Health Policy as Industrial Policy." *California Management Review*, 32: pp. 124–150, 1990.
U.S. Congress, Office of Technology Assessment. *Global Standards: Building Blocks for the Future*. Washington, D.C.: U.S. Government Printing Office, 1992.
U.S. Department of Commerce. *National Trade Estimates Report on Foreign Trade Barriers: Japan*. Washington, D.C.: U.S. Government Printing Office, 1993.
Upham, Frank. "Privatizing Regulation: The Implementation of the Large-Scale Retail Stores Law." *Political Dynamics in Contemporary Japan* (Gary D. Allinson and Yasunori Sone). Ithaca, N.Y.: Cornell University Press, pp. 264–294, 1993.
Woodall, Brian. *Japan Under Construction: Corruption, Politics, and Public Works*. Berkeley, Calif.: University of California Press, 1996.
———. *Pork Barrel Politics in Japan: Trade Friction, Public Works and the Triadic Syndicate*, 1995-1988. Doctoral dissertation, University of California at Berkeley, 1990.
Yoshikawa, Aki. "the Other Drug War: U.S.–Japan Trade in Pharmaceuticals." *California Management Review*, 31/2: pp. 76–90, 1989.

12

The Uruguay Round's Agreement on Technical Barriers to Trade: Implications for the United States and Japan

William J. Long and Kimberly Wildner

INTRODUCTION

This chapter examines the GATT's Uruguay Round Technical Barriers to Trade (TBT) Agreement and its possible effects on U.S.–Japan trade relations. The rationale for an international agreement governing technical barriers to trade is straightforward: nontariff trade barriers to trade (including technical barriers) have been steadily increasing, undermining many of the tariff-reduction gains achieved in prior GATT negotiations. In the last two decades, multilateral trade liberalization has often been compared to draining a swamp. That is, as the level of tariffs are drawn down, the submerged and stubborn stumps of nontariff barriers are revealed and must also be removed. Technical barriers to trade are increasingly among the most difficult barriers to remove, not only because of their technical/legal difficulties, but because of their domestic political purposes. Nonetheless, the benefits awarded through international cooperation in removing technical barriers to trade outweigh the gains individual nations achieve through failure to harmonize standards and practices. These benefits can include: (1) enhanced product quality and reliability at competitive prices; (2) improved health, safety, and environmental protection; (3) greater compatibility and interoperability of goods and services; and (4) increased distribution efficiency and ease of maintenance (*Introduction to ISO*).[1] In view of these benefits, the Uruguay Round negotiations of GATT included the Agreement on Technical Barriers to Trade that seeks to prohibit any government regulation "with the effect of creating unnecessary obstacles to international trade" (Agreement on Technical Barriers to Trade, Annex 1A, Article 2.2).

THE TBT AGREEMENT

The Tokyo Round of GATT was the first to significantly address nontariff barriers to trade, such as subsidies, government procurement rules, and technical barriers. The negotiation concluded in 1979 with a series of independent international agreements relating to nontariff measures, including a technical barriers agreement (the Standards Code). It is important to note that the Standards Code was an independent agreement, and was not obligatory on GATT member nations that signed the General Accords. Under the Standards Code, parties were not to discriminate against imports in cases where the exporting nation gave positive assurance that its products conformed with the host country's technical regulations and standards. The parties to the Tokyo Round also agreed, whenever possible, to accept test results and certification issued by relevant bodies in the exporting nations. The Standards Code also included commitments against product discrimination and for the use of international standards when possible.

The Standards Code lacked "teeth," however. The authority and dispute resolution mechanisms in the Standards Code were specific to the code and did not parallel the procedures under other GATT agreements. Parties were not legally obligated to carry out dispute resolution panel findings, and the lack of specific time constraints for prosecution of complaints left ample opportunity to delay the resolution process.

The new TBT Agreement extends the Standards Code by seeking to ensure that certification, accreditation, and quality system regulation procedures, as well as technical regulations and standards, do not create unnecessary obstacles to trade. As in the past, because of the link between technical standards and national policy objectives, the TBT Agreement recognizes that countries have the right to establish protection at levels they consider appropriate to maintain human, animal, or plant life, health standards, or the environment, and participating countries should not be prevented from taking measures necessary to ensure adequate levels of protection in these areas. The TBT Agreement, following the precedent set forth in the Tokyo Round Code, also encourages countries to use international standards where appropriate, but it does not require them to change their levels of protection as a result of international standardization.

The TBT Agreement's innovations include expanded coverage to processes and production methods, as well as products. The "conformity assessment procedures," in other words those used, directly or indirectly, to determine that relevant requirements in technical regulations or standards are met, are expanded to cover a broader range of processes and the disciplines made more specific. Notification provisions applying to local government and nongovernmental bodies are elaborated in more detail than in the Tokyo Round agreement, which bound only central governments and, to a lesser degree, state and provincial governments. Also, included as an annex to the TBT agreement is the Code of Good Practice for the Preparation, Adoption, and Application of Standards ("Code of Good Practice"), which is open to acceptance by private sector bodies as well as the public sector agencies. This Code of Good Practice will, among other things, require standardizing bodies to publish a notice to the World Trade Organization (WTO) of any standards under development or changes in current standards policy, and will allow an opportunity for comment by foreign parties.

Notably, the new agreement does not provide for an automatic right to gain recognition in a given country under another nation's laboratory accreditation, inspection, or quality system registration scheme. Rather, under the TBT Agreement, "members shall give positive consideration to accepting as equivalent technical regulations of other members, even if these regulations differ from their own, provided they are satisfied that these regulations adequately fulfill the objectives of their own regulations" ("Agreement on Technical Barriers to Trade," Article 2.7). The TBT Agreement, however, encourages signatories to negotiate "mutual recognition" of the results of each other's conformity assessment procedures, and advocates direct participation by conformity assessment bodies in foreign conformity assessment procedures.

To better illustrate the scope and impact of the key changes, stemming from both TBT Agreement and the (WTO), we will view each separately and in greater detail.

Nondiscrimination and Conformity Assessment

The Tokyo Round Standards Code applied the principles of national treatment and nondiscrimination only to product testing and certification programs. These principles require that the treatment of imports by a GATT member country be no less favorable than the treatment accorded to like products of domestic origin or imports from another nation. The TBT Agreement extends this basic obligation to cover a range of conformity assessment procedures as well. These include laboratory accreditation, recognition, and quality system registration programs such as ISO 9000.

Although the TBT Agreement made progress in extending conformity assessment, it provides only a limited basis to encourage acceptance of the test results or laboratory accreditation across national borders. Member nations are encouraged to give "positive consideration" to accepting equivalent technical standards when they adequately meet the objectives of their own legislation. This illustrates the vital importance of recognizing and preserving national sovereignty, which a system of automatic recognition of foreign technical standards would challenge. The agreement encourages signatory nations to harmonize conformity assessment through mutual recognition of each other's procedures. An example of the benefits afforded by mutual recognition is demonstrated by the United States and European Union (EU). In early 1994, the EU and the United States entered negotiations with a goal of producing a framework where exports from each nation were encouraged through mutual recognition of third-party product test results, inspections, and certifications. This has the benefit for the United States of increasing its access to European markets and overcoming the barriers imposed by the EU product conformity requirements. Similarly, the EU gained improved access to U.S. markets. In addition, a mutual recognition agreement will gain the EU formal U.S. government assurance that American entities are capable of performing inspection and certification services at the required level necessary to fulfill EU mandates, and vice versa. Mutual recognition agreements can facilitate increased trade by reducing costs, inefficiencies, and barriers.

Extension of Coverage to Nongovernmental Organizations

The TBT Agreement extended multilateral rules to private standards organizations, such as European regional standards developers and the American National Standards Institute (ANSI). Central governments will now be responsible for implementing the agreed rules and applying the TBT Agreement's principles at any level of government or within any private-sector body involved in the national standards system. Conversely, the Tokyo Round Standards Code bound only central governments, and less rigidly, subnational governments, to its obligations.

Further, the new Code of Good Practice provides a model for extending the agreement's rules to private standards bodies. The code outlines general principles for nongovernmental organizations in developing and applying standards. These goals include national treatment and nondiscrimination of products, publication and dissemination of work in progress, implementation of a sixty day open comment period before adopting standards, and restraint in applying standards that could act as trade barriers. The Code of Good Practice is voluntary, however, and lacks an enforcement mechanism. Nonetheless, with wide acceptance among WTO countries, the code could open a channel of communication between national standards organizations and private standards bodies, and improve dialogue and informal dispute resolution.

The next two elements discuss those changes caused, not by the TBT Agreement per se, but through association within the wider Uruguay Round accords.

Membership Expansion

Significantly, the TBT Agreement increased the number of countries bound by the agreement. As noted, the Tokyo Round Standards Code was a stand-alone agreement with its own institutional provisions and dispute settlement procedures. In other words, not all signatories to the GATT were parties to the Standards Code; only those who specifically accepted the Standards Code were bound by it. As of November 1993, there were forty-six signatories to the Tokyo Round TBT code. Most of these members were the industrialized nations, including United States and Japan.

However, with the establishment of the WTO following the successful conclusion of the Uruguay Round, membership in this new organization is available only to countries that are signatories to the GATT and who agree to adhere to all the Uruguay Round agreements. The TBT agreement, thus, is binding on all members of the WTO. The signatories to the Uruguay Round Agreement, including the new TBT Agreement, will include seventy-eight additional nations, many from the developing world. These additions will bring the total number of countries adhering to the TBT code to 124. This extension of rules to new parties is an important part of strengthening the multilateral trading system.

Dispute Settlement Mechanism

The TBT Agreement negotiated under the Uruguay Round and the Tokyo Rounds Standards Code contains different dispute settlement mechanisms. Unlike the Standards Code, the TBT Agreement's dispute mechanisms are binding on the parties. Trade actions involving technical barriers will now be considered as part of an integrated dispute resolution mechanism under the WTO. The procedure is consolidated in the Understanding on Rules and Procedures Governing the Settlement of Disputes (DSU) and applies to the entire WTO. It improves the existing system by providing strict time limits for each step in the process and for automatic enforcement of its provisions.

Under the WTO dispute settlement procedures, a member nation with a complaint about another member nation's actions first requests consultations. If the parties fail to come to an agreement within sixty days, the complaining party may request the establishment of a panel to investigate the issue and make appropriate recommendations in a report to the Dispute Settlement Body[2] (DSB). When consultations are denied, the complaining party may request an investigative panel. Panels will consist of three persons of appropriate background and experience from countries not party to the dispute. The DSU envisions that a panel will normally complete its work within six months or, in cases of urgency, within three months. During this time, the parties may voluntarily agree to follow alternative means of dispute settlement, including good offices, conciliation, mediation, or arbitration.

Once the panel report is adopted, the affected party will have to notify the DSB of its intentions with respect to implementing the adopted recommendations. The affected party will be given a reasonable period to comply with the panel's recommendations. The allotted time period is to be decided within forty-five days of adoption of the report by mutual agreement of the parties and with approval by the DSB, or within ninety days of adoption through arbitration.

Further provisions of the DSU set rules for compensating or suspending concessions in the event of non-implementation. Within a specified time frame, parties can negotiate a mutually acceptable compensation. Where negotiation is not successful, a party to the dispute may request permission to suspend concessions or other obligations to the other party. In principle, concessions should be suspended in the same sector as that in issue in the panel case. If this is not effective, the suspension can be made in a different sector of the same agreement. If this is not effective and if the circumstances are serious enough, the suspension of concessions may be made under another agreement.

In the past, a single party, including the party under investigation, had the power to block the decision to create a panel, the approval of the panel's report, and the authorization of countermeasures. The dispute settlement procedure was, in essence, voluntary. Under the Uruguay Round Agreements, the establishment of a panel and the adoption of the panel's report will be automatic, unless the DSB, by consensus, chooses not to adopt the recommendation report. Likewise, under prior arrangements, the losing party had no means of appellate review, whereas the new agreement establishes a seven-member standing appellate body. Appeals are limited to issues of law, and appellate reports, as with the original recommendation reports, are automatically adopted by the DSB unless, by consensus, they choose otherwise.

BILATERAL (U.S.–JAPAN) TECHNICAL BARRIERS ISSUES

Currently, numerous technical barriers to trade exist both in the United States and Japan. The following is a partial listing of each nation's barriers.

U.S. Barriers in Standards, Testing, Labeling, and Certification

General objections to the U.S. technical standard barriers include: (1) the complexity of the regulatory system that acts as a structural impediment to market access; (2) discrepancies with regard to international standards; (e.g., electrical currents); (3) diversity among standard setting authorities, i.e., more than two thousand, seven hundred state and municipal authorities in the United States, which require particular safety conditions for products sold or installed within their jurisdiction; and (4) high costs for foreign companies because there is no central source of information on standards and conformity assessment. Sectors generally affected by U.S. technical barriers to trade include:

- Ceramicsware and wine imports—Maximum ceilings of lead release action levels.
- Food, drug, cosmetic, and medical devices—California's Safe Drinking Water and Toxic Enforcement Act requires warning labels on all products containing substances known to cause birth defects or harm, including lead.
- Glass containers in food and beverage industry—The Public Resources Code of California set certain recycled glass composition requirements.
- Electrical Products—The National Electrical Code requires particular testing and certification.
- Industrial Fasteners—The 1990 Fastener Quality Act (FQA) requires testing and certification by an accredited laboratory regardless of ISO 9000 quality assurance.

An area of particular concern for Japan is the American Automobile Labeling Act. Enacted under Section 210 of the Passenger Motor Vehicle Content Information Disclosure Act of October 1992, the act requires all passenger cars and light trucks to label whether and to what extent the parts of the product and the product itself are of domestic origin. Specifically, labels must indicate (1) the percentage of the U.S. and Canadian parts; (2) the country, state, and city of final assembly; (3) the top two countries supplying parts and the percentages supplied by each country (if countries other than the United States and Canada supply 15 percent or more of the parts in the vehicle); (4) the engine's country of origin (the country adding 50 percent or more of the value or the most added-value); and (5) the transmission's country of origin (the country adding 50 percent or more of the value or the most added-value). The law took effect October 1, 1994, and violations are subject to a fine of $1,000 per vehicle.

Japanese Barriers in Standards, Testing, Labeling, and Certification: Concerns for the United States

Generally, Japanese standards are often criticized for their complexity and lack of transparency. Specific U.S. sectors particularly affected by Japan's use of standards, testing, and labeling as nontariff barriers include pharmaceuticals, medical devices, food additives, and fresh agricultural products (*International Trade Reporter*, p. 615). The WTO addresses these last two sectors under a separate agreement on sanitary and phytosanitary measures.

In 1992, the Japanese Government agreed to conduct a comprehensive review of its standards, certification, and testing procedures that act as technical barriers to trade. The Office of the Trade and Investment Ombudsman (OTO) conducted this review and the Japanese government promised their reply to OTO's recommendations by May 1993. Among the concerns raised by the U.S. government during this review stage were the following:

- Overtime Import Processing Fees—Extremely high fees are imposed for weekend and evening customs clearance of high-volume, low-value shipments. Estimated at $1,200 per hour, these fees greatly exceed the actual cost of customs clearance.
- Freshness Labeling—Not adopting CODEX standards for date-marking processed food products. Retailers require date of manufacturing rather than "best before" dates, which is the common method in the United States.
- Feed Grains—Strict policies and regulations in this industry restrict competition, thus inflating prices.
- High Pressure Gas Laws—Standards in this law restrict U.S. exporters' access in various sectors, including air conditioners, refrigeration equipment, supercomputers, and aircraft support equipment. Design standards, rather than performance standards, limit flexibility with regard to technological advances.
- Water Purification Systems—Japan Water Works Association uses a type approval rather than a performance standard, which limits/delays introduction of innovative products to the Japanese market.

In the initial negotiations, none of the above issues was resolved to the satisfaction of U.S. companies. Further complaints by U.S. companies concerning technical barriers included:

- Gun and Sword Law—This law requires nail and staple "guns" in the construction industry to be registered.
- Steel Structure Fire Resistant Test—Official approval certification can be obtained only in Japan.
- Imported Refrigerant Recovery and Recycling Equipment—Testing is accepted only from MITI's International Trade and Industry Inspection Institution.
- Absorption Chillers Installment in Government Buildings—A Ministry of Construction policy requires quality levels beyond standard. The United States has dominance in this industry.
- Gas Systems and Components—Used predominantly in semiconductor manufacturing equipment, this sector is restricted by a lack of clear standards and clearly defined procedures in the High Pressure Gas Law.

POTENTIAL EFFECTS OF THE TBT AGREEMENT ON U.S. AND JAPANESE BILATERAL TRADE

The implications of the Uruguay Round Agreements and the TBT Agreement to U.S.–Japanese trade are potentially significant, but uncertain. While the TBT Agreement made a great deal of progress from the Tokyo Round Code, numerous provisions remain largely voluntary. This leaves the potential for actual gains dependent on each nation's level of independent compliance within the agreement's flexible parameters.

The transparency[3] and non-discrimination requirements for issuing product approval and the amended definitions of technical regulations and standards to cover processes and production methods that could potentially restrict imports will have an impact on both U.S. and Japanese regulations. Key examples include Japan's High Pressure Gas Law, which may come under scrutiny for its complex approval process and lack of transparency. Likewise, the American Automobile Labeling Act could be investigated for protectionist intent in violation of non-discrimination requirements. Thus, in the near term, the severity of these trade issues between the United States and Japan may increase, merely due to the availability of agreed-upon international definition and compliance measures. As each country resolves the inconsistencies and complexities in their respective approval processes, however, the number of possible trade disputes in technical barriers are likely to fall. Also, there is a general opinion[4] that since neither United States nor Japanese standards discriminate based on the origin of imports (with a few exceptions), the agreement will allow the two countries to essentially continue as they have been.

On the other hand, the agreement's vague, nonbinding language could result in little change. The new TBT Agreement seeks to ensure that technical regulations and standards, as well as certification, accreditation, and quality systems regulation, are not more trade restrictive than necessary to fulfill a legitimate government objective. Yet, this leaves the questions of what is "legitimate" and what is "not more restrictive than necessary" open to debate. Furthermore, regulations are not required to be proportional to the economic costs, but are ambiguously set by historical precedent.

With regard to harmonizing standards, the TBT only encourages the use and development of international standards for technical regulations. This issue has often been the source of complaints against the United States, particularly in regards to electrical products. Forces outside the TBT Agreement may move the United States toward international standards, however. Over time, as international trade and regional trade blocs continue to expand, the United States will undoubtedly shift standard formulation to a national organization operating under international standards and the Code of Good Practice. Further incentive for U.S. cooperation in international standards may come from the amount of recent attention that has been paid to the area of information technology. This sector currently faces competitive disadvantages from inadequate participation by the United States in international standardization. This illustrates the possibility that private bodies, operating under the Code of Good Practice, could drive the standards formulation process.

The movement toward reciprocity in conformity assessment procedures leads to another area of uncertainty with the new TBT Agreement. On this point the U.S. National Research Council concludes:

Only experience will resolve serious questions as to whether the new agreement can provide a basis to challenge conformity procedures that constitute trade barriers. It is likely, moreover, that problems of interpretation will arise in areas such as laboratory accreditation and quality systems registration in environmental management. It is uncertain, for example, how national governments will interpret Article 6 of the agreement, which requires "whenever possible, that the results of conformity assessment procedures in other Members are accepted, even when those procedures differ from their own, provided they are satisfied that those procedures offer an assurance of conformity." (National Research Council, 1995, p. 120)

Thus, the degree to which actual changes may occur will depend on each member's interpretation and level of compliance. Further progress will also depend on the evolution of national cooperation and the multilateral precedents that will be set by the WTO's enforcement mechanisms.

The establishment of the World Trade Organization and the changes in dispute settlement processes should improve trade conflict resolution, particularly in terms of speed and reliability of adjudication. Under the Tokyo Round, no technical barriers to trade cases were formally brought before a panel for final decision. Much of this is attributable to the separation of the code from the GATT dispute resolution process. Under the WTO, there are already two standards-related cases under review (neither involving the United States and Japan). Both nations have historically maintained U.S.–Japanese bilateral dispute resolution outside of the GATT channels, and will most likely continue to do so. We can expect both nations to continue a conservative policy of choosing only cases strongly demonstrating the agreement's principles.

The strengthening of the dispute settlement's procedure could raise U.S. concerns that its unilateral use of its trade laws (e.g., Section 301) would be curtailed and that other WTO members could challenge a multitude of other U.S. laws, such as those designed to protect safety, health, and the environment. The United States may have to admit to greater restraints in its trade practices, yet more efficient opportunities are available in standards harmonization and mutual recognition negotiation. However, studies by the General Accounting Office on the effect of the settlement procedures on U.S. laws and regulations have shown that the United States will still be able to use its trade laws and other domestic policies without interference (United States General Accounting Office, GAO/GGD-94-83a, July 1994, p. 4).[5]

Japanese officials are concerned about the possibility of multilaterally sanctioned retaliation under the new measures. As one Japanese study reported, the new arrangements allow the party with a complaint more scope for conclusive action and invite "a flood of requests for dispute settlement under the WTO" (*1994 Report on Unfair Trade Policies by Major Trading Partners: Trade Policies and WTO*, p. 36). On the other hand, there is a benefit to Japan and those members who have been damaged by unilateral actions. The DSU will also act as a means to fight back against unilateral trade actions, thus possibly acting as an effective restraint against nations imposing these actions indiscriminately. Initial indications suggest that the Japanese may rely upon WTO dispute settlement procedures to correct such actions. This includes any 301 sanctions taken by the United States against Japan that may be in violation of the WTO Agreement. In this case, "Japan should refer the matter immediately to the GATT or DSB under WTO" (Industrial Structure Council, 1994, p. 38). However, Japan must balance the advantages of greater clarity in global standards and the availability of recourse in a multilateral forum, with the inevitable demands and intrusions into its sovereignty that the new agreement will entail.

Japanese export promotion policies may benefit from the expanded number of participants to the TBT Agreement. Japan has been active for many years in combining foreign development assistance with the recipient's adoption of standards-setting programs. This offers both nations the added benefit of not only expanding markets and trade, but a market with compatible standards as well. In fact, the Japanese Five-Year Plan for Industrial Standards calls particular attention to the role that this type of technical cooperation can play in expanding markets and adjusts Japanese national strategy accordingly (Office of Technology Assessment, TCT-512, p. 90).[6] For Japan, the WTO and the TBT Agreement offers potential to expand market share by fostering standards convergence. The United States has no equivalent program because of its lack of a comprehensive national standard policy.

CONCLUSION

As part of a larger effort to reduce nontariff barriers to trade, the most recent round of GATT negotiations has expanded the scope and rigor of, and adherents to, its technical barriers to trade provisions. This chapter discusses some of the implications involved for the United States and Japan. The expansion of the TBT Agreement and the formation of the WTO have made undeniable progress in encouraging fair and efficient uses of technical standards in international trade, but the full impact of the agreement depends critically on the continued effort of member countries and the WTO. As trade continues to expand as a principle source of international economic growth, and reliance on international standards and measures continues to grow, the United States and Japan must adjust their national policies to maintain their competitiveness and continue as economic leaders. This effort will be particularly challenging for the United States, considering its long history of voluntary standards formation. Japan too, faces challenges in reducing its reliance on technical barriers to trade and in opening its markets.

NOTES

1. A press release from the International Organization for Standardization (ISO).
2. The Dispute Settlement Body acts in lieu of the WTO General Council to administer all dispute settlement procedures. It is composed of representatives from all 124 signatory nations.
3. In this case, transparency refers to the extent to which laws, regulations, agreements, and practices affecting international trade are open, clear, measurable, and verifiable.
4. We would like to thank Suzanne Troje, director of Technical Trade Barriers (USTR), for her assistance and insight into our research.
5. For a more detailed review of this study's findings, consult GAO/GGD-94-83b, Volume II of the above cite, pp. 40-50.
6. Examples include the Philippines, where the Japanese International Cooperation Agency sent a thirteen person team to conduct a five hundred person a day study of the Philippine national standardization system and provided a $23.1 million grant to establish three regional labs. In addition, each year the Japanese government pays for thirty-two people from developing nations to come to Japan to engage in language and technical training.

REFERENCES

GATT Secretariat. "The Final Act of the Uruguay Round: A Summary." *International Trade FORUM*, January 1994, pp. 10, 19–20.

Goddin, Scott R. "Standards Code: Agreement on Technical Barriers to Trade." *Business America*. January 1994, pp. 17–18.

National Research Council. *Standards, Conformity Assessment, and Trade: Into the 21st Century*. Washington, D.C.: National Academy Press, 1995, pp.103–48.

Nelson, Kathryn. "International Dispute Resolution: How NAFTA and GATT Can Help," *Review of International Business & Law* 1, no. 3 (fall 1994): 9.

1994 Report on Unfair Trade Policies by Major Trading Partners: Trade Policies and WTO. Tokyo, Japan: Industrial Structure Council, 1994.

"1994 Report on U.S. Barriers to Trade and Investment." Doc No I/194/94. Services of the European Commission. April 1994, pp. 55–60.

"1994 National Trade Estimate Report on Foreign Trade Barriers." United States Trade Representative, 1994. pp. 149–53.

U.S. Congress, Office of Technology Assessment. *Global Standards: Building Blocks for the Future*. TCT-512. Washington, D.C.: U.S. Government Printing Office, March 1992.

U.S. Congress, Senate. "Overview of the Results of the Uruguay Round." Hearing before the Committee on Commerce, Science, and Transportation. 103rd Congress, 2nd Session, June 16, 1994.

U.S. General Accounting Office. "The General Agreement on Tariffs and Trade: Uruguay Round Final Act Should Produce Overall U.S. Economic Gains." GAO/GGD-94-83a/b, July 1994.

Glossary of Japanese Terms

Amakudari	"Descent from heaven"
Gaiatsu	Foreign critisms and pressures
Itaku kenkyu	Contract research
Kanji	Japanese characters
Keidanren	Federation of Economic Organizations
Shingikai	Advisory councils
Kohsetsushi	An acronym for *koh*, public; *setsuritsu*, establishment; and *shikenjo*, testing laboratory
Gyosei Shido	Administrative guidance

Index

Administrative guidance (*gyosei shido*), 41
Advanced Mobile Phone Service (AMPS), 88, 89, 90
Advanced Technology Program (ATP), 17
Agreement on Technical Barriers to Trade (TBT). *See* Tokyo Round TBT Agreement; Uruguay Round TBT Agreement
American College of Radiology/National Electrical Manufacturers Association (ACR/NEMA), digitalized medical imaging standard, 98, 105
American National Standards Institute (ANSI), 138–39
APEC. *See* Asia Pacific Economic Cooperation
Asia
   international standards and conformity practices in, 123
   registrations to ISO 9000 in, 129
   standards bodies in, 86
Asia Pacific Economic Cooperation (APEC)
   Bogor Declaration, 136
   and developing countries' standards/infrastructure needs, 137–38
   standards and conformance assessment agenda, 123, 125–27, 139–40
   strategy to 2020, 125, 136–41
   and TBT implementation, 137
Australia, R&D tax incentives in, 14–15

Bangemann, Martin, 122

CAD/CAM software, 83–84
Cellular communications equipment standards, 87–92
   of analog systems, 88–90, 91–92
   of digital systems, 90–91, 92
Conformity assessment
   as barrier to trade, 69, 123–24
   and mutual recognition agreements, 120, 135–36
   need for international harmonization of, 70–71, 134
   procedures, 132
   testing laboratories, 133–34
   and Uruguay Round TBT agreement, 128–30, 131–32. *See also* International Standards Organization (ISO)

DDI (Japanese firm), 91
Developing countries
  as signatories to Uruguay Round
    Agreements, 128
  infrastructure and standards systems
    needs of, 137–39
DICOM (Digital Imaging and
  Communications in Medicine)
  standard, 98–99, 100, 102, 103,
  104, 105
Discriminatory practices
  in certification requirements, 126
  in Japanese standard-setting, 123–24
  and Uruguay Round TBT agreement,
    131–32

Engineering workstation market, 83
Environmental standards, 121, 126
European Community (EC), Framework
  programs, 13, 17, 20–21, 150
European Union
  and international standards, xxi–xxii,
    121–22
  MRA negotiations with U.S., 122,
    126–27, 135–36, 139

Fiscal Investment and Loan Program
  (FILP), 18, 19
Fuji Photo, 96
Fujitsu, 59–60, 81

General Agreement on Tariffs and Trade
  (GATT), adoption of, 119–20. *See
  also* Tokyo Round TBT
  Agreement; Uruguay Round TBT
  Agreement
Generic technology, 7
Global System for Mobile
  Communications (GSM), 90

Hitachi, 56–58, 60, 61, 94, 98

IDO (Japanese firm), 91–92
Industrial Science and Technology
  Frontier Program, 15–16, 17–18
Information technology

  size of Asian market in, 85
  U.S.–Japan trade friction in, 77–86.
    See also Japanese information
    technology standards
Infrastructure, definitions of, 23n.1
Infratechnologies, 8, 9
Intellectual property rights, 17, 87, 90
International Electrotechnical Committee,
  69–70, 72
International Standards Organization
  (ISO), 69–70, 72, 138
  ISO 9000 standards, 7–8, 81–83, 121,
    129, 132
  ISO 14000, 121
IS&C (Image Save and Carry), 93, 99–
  105
ISO. *See* International Standards
  Organization

Japan Bicycle Development Association,
  36–37
Japanese digitalized medical imaging,
  93–106
  CR (computed radiography), 94–96
  CT (computed tomography), 93–94,
    96
  DICOM-based Standard II in, 102,
    103, 104–6
  international standards for, 98–99
  IS&C-based Standard I in, 99–105
  MRI (magnetic resonance imaging),
    93–94, 96
Japanese electronics industry, 52–63
  basic research approaches in, 54–63
  competitive advantage of, 113–14
  and component miniaturization
    research, 52–54
  and greater systematization of
    products, 52–54
Japanese electronics packaging, 107–14
  competitive advantage in, 109–11
  supporting infrastructure in, 111–12
Japanese hospital information systems
  (HIS), 97
Japanese Industrial Standards (JIS), 68,
  75, 80, 124, 151
Japanese industry

Index   179

competitive advantage of, 109–11,
    113–14
dominant strategic logic of, 47–48
implications of Uruguay Round
    agreements for, 170–73
market share–driven strategy of, 48–
    49
as model of economic success, 77–79
product proliferation in, 49–50
standards assistance to developing
    countries by, 138
technical trade barriers of, 84–85,
    169–70. *See also* Research and
    development, Japanese
Japanese information technology
    standards, 77–86
adoption of ISO 9000 series, 81–82
as barriers to U.S. operations in Japan,
    84–85
foreign involvement in, 80
and intercompatibility, 80–81
and U.S.–Japanese MOSS Talks, 80,
    91, 150
Japanese regional technology projects,
    39–45
Kohsetsushi centers, 35–38, 40–45
Technopolis program, 39
third sector approach in, 39–40, 43–
    44, 46n.15
Japanese small- and medium-sized
    manufacturing enterprises (SMEs),
    27–45
development of new products and
    strategies in, 30–31
dualistic view of, 27–28
effects of economic recession on, 27,
    31
institutional support for, 32–33, 35–38
and Japan's modernization policies,
    31–35
regional technology
    projects/partnerships, 28, 33, 39–
    45
sector features, 30
and vertical linkages with large
    companies, 27–28, 30
Japanese Technology Evaluation Center
    (JTEC), 107, 110
Japan Small Business Corporation, 32, 36

Kajiwara, Ken'ichiro, 103
Kelly, Michael J., 107
Key Technology Center (KTC), 19–20
Kohsetsushi centers, 35–38, 40–45
    administration and funding of, 36–37
    MITI guidance of, 41
    weakness of research and technology
        quality in, 41–42

Large-Scale Project, 15
Latin America, and international
    standards, 121
Long, William J., 163

Market failure
    as rationale for government R&D
        support, 10–13
    responses of industrialized nations to,
        13–22
Market-Oriented Sector-Selective
    (MOSS) agreement, 80, 91, 150
Medical and Welfare Equipment
    Technology Project, 15
Medical Information System
    Development Center (MEDIS), 99–
    100, 101–2, 104, 105
Methé, David T., 48
Minato, Kotaro, 103
Ministry of Health and Welfare (MHW),
    147
Ministry of International Trade and
    Industry (MITI), 15–16, 17, 18, 19,
    36, 39, 41, 112, 124, 147
MIPS (Medical Imaging and Processing
    Systems) standards, 98
Mitsubishi Electric (MELCO), 60, 61
Mitsutoyo, Inc., 83
Morgan, James, 85
Mossbacher, Robert, 122
Motorola, 88, 90, 91
Murata, 113
Mutual Recognition Agreements (MRAs),
    120
    U.S.–European Union negotiations,

122–27, 135–36, 139

National Electrical Manufacturers Association (NEMA), 98, 99
National Voluntary Conformity Assessment Systems Evaluation (NVCASE), 122
NEC (Japanese firm), 50–51, 58, 60, 61, 80, 82, 90, 98
New Energy and Industrial Development Organization (NEDO), 16
Next Generation Project, 15
Nippon Telephone and Telegraph (NTT), 39, 54–55, 149
Nippondenso, 113
NIST. *See* U.S. National Institute for Standards and Technology
North American Free Trade Agreement (NAFTA), 120–21
NTT. *See* Nippon Telephone and Telegraph

Ozawa, Ichiro, 150

Picture Archiving and Communication Systems (PACS), 97–98
Publicly Available Standards, 72
Puffert, Douglas J., 87

Research and development (R&D):
  knowledge exploration versus knowledge exploitation in, 48–49, 61–63
  public-sector involvement in, 4–13, 18
  rationales for, 10–13
  relationship between private- and public-sector R&D, 4–5
  tax incentives for, 14
Research and development, European:
  ESPRIT program, 21
  EUREKA program, 13–14, 21, 22
  Frameworks programs, 13, 17, 20–21, 150
Research and development, Japanese:
  in early-phase (generic) technology, 15–20
  energy and environmental, 16

funding of, 15–20, 24nn.10, 17
and government industrial research institutes (GIRIs), 19
government involvement in, 112–13
and intellectual property rights, 17
joint ventures in, 20
knowledge exploitation in, 48–49, 61–63
in next generation technologies, 16–17
Research and development, U.S.:
  ad hoc projects, 22
  conformity assessment in, 122, 133–34
  funding of, 15, 23n.3
  government involvement in, 112–13
  and R&E (research and experimentation) tax credit, 14

Scandinavia, Nordic Mobile Telephone (NMT) system standard, 88–89
Science and Technology Agency (STA), 18
Semiconductor technolology, 17, 77
Shapira, Philip P., 27
Small and Medium Enterprise Agency (SMEA), 32
Sony, 113
Standards:
  benefits of, 125
  compatibility, and competitive strategy, xix–xxi, 87–88
  de facto, 71–72, 73
  defined, xviii, 145–46
  development process, xviii–xxii
  discriminatory, 146, 153
  globalization of, 69–73
  mandatory, 146
  promotional, 155–56, 159n.8
  purposes of, 68–69
  and rapid technological change, 71–72
  and trade friction, 145–46
  and U.S.–Japanese business competition, 83. *See also* Conformity assessment
Standard-setting process, international, xxi
Standards, European, xxi–xxii

Index

and Japanese industry, 81, 82
provisional standard system, 72
Standards, Japanese
and conformity assessment, 123–24
as discriminatory, 123–24
distinctive cultural features of, 79
and Japan's dual political economy, 156–57
and medical/pharmaceutical products, 148–49
in procurement criteria, 149
proliferation of, 151
as promotional, 155–56
as protective, 153–55
in safety and testing standards, 147–48
standards-setting procedure, 151–53
and U.S.–Japan trade friction, 149–51
Stern, John P., 75

Tanabe, Koji, 68
Tassey, Gregory, 3
Technical barriers to trade
conformity assessment as, 69, 123–24
and discriminatory certification requirements, 126
and free trade agreements, 120–21
and international harmonization of standards, 121–24
by Japan, 84–85, 169–70
standards as, 119–20
and Tokyo Round TBT Agreement, 70, 127–28, 129, 130, 164, 165, 166. *See also* Uruguay Round TBT Agreement
Technological infrastructure
conceptual model of, 21
elements of, 7–8
and "mixed economy" model, 22
regional strategies in, 13–14
roles and economic impacts of, 3–9
tax incentives as strategy in, 14–15
Technopolis program, 39
Time-Division Multiple Access (TDMA), 90, 91
Tokyo Round TBT Agreement, "the Standards Code," 70, 127–28, 129, 130, 164, 165, 166

Toshiba, 58–59, 103
Toshiba Medical, 94, 98
Total Access Communication System (TACS), 88, 89, 90, 91, 92
Total quality management (TQM), 7

United States
and European Union, MRA negotiations with 122, 126–27, 135–36, 139
implications of Uruguay Round agreements for, 170–73
and international standards, 122
small- and medium-sized manufacturing enterprises, 30
standards assistance to developing countries by, 138–39
technical trade barriers of, 165–70. *See also* Research and development, U.S.
Uruguay Round TBT Agreement, 127–32, 163–73
conformity and nondiscrimination assessment in, 128–30, 131–32, 165–66
dispute settlement mechanism in, 130–31, 167–68
expanded signatories and scope of, 127–28, 164–65, 166–67
extension to non-governmental bodies and local governments, 130, 166
implementation of, 137
and international standards, xxi, 70
and NAFTA, 120
potential effects on U.S.–Japan bilateral trade, 170–73
principle of "national treatment" in, 120
U.S. Advanced Technology Program (ATP), 15
U.S. Federal Communications Commission (FCC), single system standard of, 88
U.S. National Institute for Standards and Technology (NIST), 15, 122, 134
U.S. Trade Representative (USTR), 91, 92

Vertical integration, 21, 27–28, 30
VLSI Project, 16, 17

Warshaw, Stanley I., 119
Wildner, Kimberly, 163
Wilson, John Sullivan, 125
Woodall, Brian, 145

World Trade Organization (WTO), 69–70, 120, 129, 130, 131

Yamashita, Isao, 80
Yeutter, Clayton, 150
Yokogawa Medical (GE), 94
Yoshikawa, 93

About the Editor and Contributors

MICHAEL KELLY, D.Sc. Currently at Georgia Institute of Technology, where he is involved in the development of educational programs in management of technology and global corporate enterprise, he was previously the director of the Defense Manufacturing Office at the Defense Advanced Research Projects Agency, and Executive Director of the National Advisory Committee on Semiconductors.

WILLIAM J. LONG, J.D., Ph.D., Professor of International Affairs and Associate Director of the School of International Affairs at Georgia Institute of Technology, is the author of *US Export Control Policy* and many articles and book chapters on international trade and technology policy. Dr. Long previously practiced international trade law in Washington, DC.

JOHN R. MCINTYRE is Director of the federally funded Georgia Tech Center for International Business Education and Research (CIBER), a national resource center, and a professor of international business management and international relations with joint appointments in the Dupree School of Management and School of International Affairs of the Ivan Allen College, Georgia Institute of Technology, Atlanta, Georgia. He is also a senior fellow of the East-West Trade Policy Center of the University of Georgia, Athens, Georgia.

He received his graduate education at McGill, Strasbourg and Northeastern Universities, obtaining his Ph.D. at the University of Georgia. Prior to joining Georgia Tech in September 1981, he was Research Associate for International Management at the Dean Rusk Center. He has had work experience with multinational firms in the U.K. and Italy. He has published articles in journals such as *Osteuropa Wirtschaft* (Munich), *Technology and Society, Public Admini-*

stration Quarterly, International Management Review, Defence Analysis (London), *Studies in Comparative and International Development, Crossroads, The Journal of European Marketing, Jeune Afrique, Le Moci* (Paris), *Politique Internationale* (Paris), *Fordham International Law Journal, International Trade Journal* as well as many book chapters. He is author and co-editor of the following books: *Uncertainty in Business-Government Relations: The Dynamics of International Trade Policy, The Political Economy of Technology Transfer, National Security and Technology Transfer: The Strategic Dimensions of East-West Trade, International Space Policy: Legal, Economic, and Strategic Options for the Twentieth Century and Beyond.* His non-academic activities include international business strategy consulting with several southeastern U.S. and European firms. He has served as a member of the Delegation of the European Communities Commission "European Union Task Force of Experts" in the U.S. since 1988. He has been editor of the annual Japanese investment yearbook in the Southeast U.S. of The Japan-America Society Inc. since 1990. He is a regular radio and TV commentator for Canadian Broadcasting Company and a correspondent for the Paris quarterly *Politique Internationale.*

DAVID METHÉ, Ph.D., Associate Professor of Corporate Strategy, University of Michigan at Ann Arbor, is currently engaged in research related to comparisons of U.S. and Japanese approaches to technology management and product development in high technology related products. He is the author of *Technological Competition in Global Industries.*

DOUGLAS J. PUFFERT, Ph.D., International Trade Analyst at the U.S. International Trade Commission, Washington, D.C., was previously an economics professor at the University of Connecticut and a consultant for The World Bank.

PHILIP SHAPIRA, Ph.D., Associate Professor of Public Policy, Georgia Institute of Technology, teaches and conducts research on industrial modernization, technology policy, and comparative economic and regional development. He is co-editor of *Planning for Cities and Regions in Japan,* and editor of *The R&D Workers: Managing Innovation in Britain, Germany, Japan and the United States.*

JOHN P. STERN, Vice President, Asian Operations, American Electronics Association, Tokyo, Japan, is the first non-Japanese to be elected a director of The Telecommunication Technology Committee. He was previously an advisor for Japanese electronics companies. He has published in the fields of Japanese commercial dispute resolution law and Japanese R&D tax policy.

KOJI TANABE is Director for International Standards Affairs, Standards Department, Ministry of International Trade and Industry (MITI), Tokyo, Japan.

He is a trade and standards expert and has been posted in Southeast Asia prior to assuming his position in Tokyo.

GREGORY TASSEY, Ph.D., Senior Economist, National Institute of Standards and Technology, Bethesda, Maryland, has published extensively on cooperative technology policy and growth, his most recent book being *Technology Infrastructure and Competitive Position.*

STANLEY WARSHAW, P.E., Ph.D., Senior Policy Advisor for Standards and Technology, U.S. Department of Commerce, was responsible for standards activities at the National Institute of Standards and Technology (NIST) from 1981 to 1994. Dr. Warshaw was a senior research scientist at Raytheon Company and American Standard.

KIMBERLY WILDNER, a Georgia Institute of Technology graduate, is currently pursuing a Ph.D. in Economics at the University of North Carolina, Chapel Hill.

JOHN SULLIVAN WILSON, Senior Staff Officer, National Academy of Sciences has researched on standards policy, and was a fellow at the Institute for International Economics in Washington, D.C. His most recent book is entitled *Standards, Conformity Assessment and US Trade: Into the 21st Century.*

BRIAN WOODALL, Ph.D., Assistant Professor of International Affairs, Georgia Institute of Technology, was an Abe Fellow, affiliated with the University of Tokyo, held a faculty appointment at Harvard University and the University of California at Irvine, and has authored several articles and books on Japan's domestic politics, trade, industrial and foreign policies. His most recent publication is *Japan's Changing World Role: Emerging Leader or Perpetual Follower?*

AKI YOSHIKAWA, Ph.D., Associate Director of the Comparative Health Care Policy Research Project, Stanford University, is the recipient of a 1995 Abe Fellowship. He taught economics at the University of California at Berkeley, and was also a Research Director at the Berkeley Roundtable on the International Economy. He is an academic consultant and has conducted policy analysis for the U.S. Office of Technology Assessment. He is the author of several books, including *Japan's Health System: Efficiency and Effectiveness in Universal Care.*